2013—2025年国家辞书编纂出版规划项目

英汉信息技术系列辞书

总主编 白英彩

A CONCISE ENGLISH-CHINESE DICTIONARY OF INTELLIGENT ROBOTICS

英汉智能机器人技术简明词典

主　编　陈卫东

主　审　蔡鹤皋　俞　涛

副主编　王贺升　王景川

　　　　卢俊国　刘丽兰

内容提要

本词典为“英汉信息技术系列辞书”之一，收录了智能机器人技术及其产业的理论研究、开发应用、工程管理等方面的词条 3 800 余条，并按英文词汇首字母顺序排列；对所收录的词条进行疏理、规范和审定。本词典可供智能系统与智能机器人相关领域进行研究、开发和应用的人员、信息技术书刊编辑和文献译摘人员使用，也适合上述专业的高等院校师生参考。

图书在版编目(CIP)数据

英汉智能机器人技术简明词典／陈卫东主编．—上海：上海交通大学出版社，2022.7
ISBN 978-7-313-24480-2

Ⅰ．①英… Ⅱ．①陈… Ⅲ．①智能机器人—对照词典—英、汉 Ⅳ．①TP242.6-61

中国版本图书馆 CIP 数据核字(2020)第 261117 号

英汉智能机器人技术简明词典
YINGHAN ZHINENG JIQIREN JISHU JIANMING CIDIAN

主　　编：陈卫东
出版发行：上海交通大学出版社　　地　　址：上海市番禺路 951 号
邮政编码：200030　　电　　话：021-64071208
印　　制：苏州市越洋印刷有限公司　　经　　销：全国新华书店
开　　本：880 mm×1230 mm　1/32　　印　　张：3.75
字　　数：108 千字
版　　次：2022 年 7 月第 1 版　　印　　次：2022 年 7 月第 1 次印刷
书　　号：ISBN 978-7-313-24480-2
定　　价：98.00 元

英汉信息技术系列辞书顾问委员会

英汉信息技术系列辞书编纂委员会

《英汉智能机器人技术简明词典》

编 委 会

序

信息技术(IT)这个词如今已广为人们知晓，它通常涵盖计算机技术、通信(含移动通信)技术、广播电视技术、以集成电路(IC)为核心的微电子技术和自动化领域中的人工智能(AI)、神经网络、模糊控制和智能机器人，以及信息论和信息安全等技术。

近20多年来，信息技术及其产业的发展十分迅猛。20世纪90年代初，由信息高速公路掀起的IT浪潮以来，信息技术及其产业的发展一浪高过一浪，因特网(互联网)得到了广泛的应用。如今，移动互联网的发展势头已经超过前者。这期间还涌现出了电子商务、商务智能(BI)、对等网络(P2P)、无线传感网(WSN)、社交网络、网格计算、云计算、物联网和语义网等新技术。与此同时，开源软件、开放数据、普适计算、数字地球和智慧地球等新概念又一个接踵一个而至，令人应接不暇。正是由于信息技术如此高速的发展，我们的社会开始迈入“新信息时代”，迎接“大数据”的曙光和严峻挑战。

如今信息技术，特别是“互联网+”已经渗透到国民经济的各个领域，也贯穿到我们日常生活之中，可以说信息技术无处不在。不管是发达国家还是发展中国家，人们之间都要互相交流，互相促进，缩小数字鸿沟。

上述情形映射到信息技术领域是：每年都涌现出数千个新名词、术语，且多源于英语。编纂委认为对这些新的英文名词、术语及时地给出恰当的译名并加以确切、精准的理解和诠释是很有意义的。这项工作关系到IT界的国际交流和大陆与港、澳、台之间的沟通。这种交流不限于学术界，更广泛地涉及IT产业界及其相关的商贸活动。更重要的是，这项工作还是IT技术及其产业标准化的基础。

编纂委正是基于这种认识，特组织众多专家、学者编写《英汉信息技术大辞典》《英汉计算机网络辞典》《英汉计算机通信辞典》《英汉信息安全技术辞典》《英汉三网融合技术辞典》《英汉人工智能辞典》《英汉建筑智能化技术辞典》《英汉智能机器人技术辞典》《英汉智能交通技术辞典》《英汉云计算·物联网·大数据辞典》《英汉多媒体技术辞典》和《英汉微电子技术辞典》，以及与这些《辞典》(每个词汇均带有释文)相应的《简明词典》(每个词汇仅有中译名而不带有释文)共24册，陆续付梓。我们希望这些书的出版对促进IT的发展有所裨益。

这里应当说明的是编写这套书籍的队伍从2004年着手，历时10年完成，与时俱进的辛勤耕耘，终得硕果。他们早在20世纪80年代中期就关注这方面的工作并先后出版了《英汉计算机技术大辞典》(获得中国第十一届图书奖)及其类似的书籍，参编人数一直持续逾百人。虽然参编人数众多，又有些经验积累，但面对IT技术及其产业化如此高速发展，相应出现的新名词、术语之多，尤令人感到来不及收集、斟酌、理解和编纂之虞。如今推出的这套辞书不免有疏漏和欠妥之处，请读者不吝指正。

这里，编纂委尤其要对众多老专家的执着与辛勤耕耘表示由衷的敬意，没有他们对事业的热爱，没有他们默默奉献的精神，没有他们追求卓越的努力，是不可能成就这一丰硕成果的。

在"英汉信息技术系列辞书"编辑、印刷、发行各个环节都得到上海交通大学出版社大力支持。尤其值得我们欣慰的是由上海交通大学和编纂委共同聘请的15位院士和多位专家所组成的顾问委员会对这项工作自始至终给予高度关注、亲切鼓励和具体指导，在此也向各位资深专家表示诚挚谢意！

编纂委真诚希望对这项工作有兴趣的专业人士给予支持、帮助并欢迎加盟，共同推动该工程早日竣工，更臻完善。

英汉信息技术系列辞书编纂委员会

名誉主任：吴启迪

2015年5月18日

前　　言

机器人学是一门涉及控制、机械、电子、材料、能源、仿生、计算机、人工智能等多个领域的交叉学科，而智能机器人技术作为机器人技术的第三代技术，代表着机器人技术的最新发展方向。智能机器人具有高度的自主性和适应性，具有多种环境感知功能，可以进行复杂的逻辑思维、判断决策、规划学习、合作协作、人机交互，并且可以在物理世界中完成各种复杂困难的任务。2015 年 5 月 19 日，国务院正式颁布《中国制造 2025》计划，文件中指出，“围绕汽车、机械、电子、危险品制造、国防军工、化工、轻工等工业机器人、特种机器人，以及医疗健康、家庭服务、教育娱乐等服务机器人应用需求，积极研发新产品，促进机器人标准化、模块化发展，扩大市场应用。突破机器人本体、减速器、伺服电机、控制器、传感器与驱动器等关键零部件及系统集成设计制造等技术瓶颈”。为了适应智能机器人技术发展的实际需求，我们编纂出版了《英汉智能机器人技术简明词典》。

在学习、研究和解决智能机器人技术的关键问题时，我们经常会遇到大量缩略语、术语、专有名词等。它们一般由专业技术词汇、工程词汇、协议名词等组成，不利于理解和记忆，这些问题会给技术人员的国内外交流和文章的写作带来不便，也给大家的学习和工作造成困难，而通过这本辞典可以有效解决这些问题。

本词典编入了智能机器人领域等相关的规范化名词术语 3 800 余条，内容涵盖了智能机器人领域科研、应用和管理等方面的内容。

参与本词典编写的人员有60余人，除了已署名者以外，还包括以下人员：王怡文、王韵清、方宇凡、田伟、付向宇、朱佳伟、朱航炜、刘成志、刘雨霆、刘哲、沈逸、张钟元、张香利、陈思恒、郎若晨、郑刘坡、郑栋梁、赵文锐、赵明、赵亮、胡辰、胡晓伟、顾景琛、倪杭、徐昊、徐璠、曹瑾、崔磊磊、章闻曦、梁佳欣、彭伟、蒋舵凡、蓝功文、魏之暄、解杨敏、宋韬、张泉、贾文川、李龙、鲍晟、刘颖、唐律邦、吴国经、廖业国、刘怡伶、胥敬文、赵小文、赵为、刘汐、刘卫平、宋博文、尚天祥、于翔宇、胡寒江、王立赛、王昱欣、周坤、顾晨岚、沈成洋、吴锐凯、张旭、郝天、谢洪乐、王光明。特别感谢席裕庚、黄迪山、郭帅、李朝东四位教授对本书初稿的审读意见和修订建议。

初稿形成后，由蔡鹤皋院士和俞涛教授主审，他们一丝不苟，精益求精，逐条审核了全书。英汉信息技术系列辞书总主编白英彩教授对本词典的编纂给予了细致而具体的指导。这里谨向各位同仁、专家致以诚挚的谢意！

由于机器人技术、人工智能技术及其产业发展迅速，新的名词、术语不断涌现，书中存在的疏漏之处，恳请各位同仁与读者不吝指教。

感谢深圳市普联技术公司董事长赵建军先生对本书出版给予的鼎力资助。

陈卫东　谨识

2022年1月

凡　　例

1. 本词典按英文字母顺序排列，不考虑字母大小写。数字和希腊字母另排。专用符号(空格、圆点、连字符等)不参与排序。
2. 词汇及其对应的中文译名用粗体。一个英文词汇有几个译名时，可根据彼此意义的远近用分号或者逗号隔开。
3. 圆括号“(　)”内的内容表示解释或者可以略去。如“connecting hardware 连接(硬)件[连接器(硬)件]”；也可表示某个词汇的缩略语，如 above ground structure(AGS)。
4. 方括号“[　]”内的内容表示可以替换紧挨方括号的字词。如“cabling 布线[缆]”。
5. 单页码上的书眉为本页最后一个英文词汇的第一个单词；双页码上的书眉为本页英文词汇的第一个单词。
6. 英文名词、术语的译名以全国科学技术委员会名词审定委员会发布的为主要依据，对于已经习惯的译法也作了适当反映，如“disk”采用“光碟”为第一译名，“光盘”为第二译名。
7. 本词典中出现的计量单位大部分采用我国法定计量单位，但考虑到读者查阅英文技术资料的方便，保留了少量英制单位。

目　录

A ………………………………………………………… 1
B ………………………………………………………… 10
C ………………………………………………………… 13
D ………………………………………………………… 21
E ………………………………………………………… 26
F ………………………………………………………… 30
G ………………………………………………………… 34
H ………………………………………………………… 38
I ………………………………………………………… 41
J ………………………………………………………… 46
K ………………………………………………………… 47
L ………………………………………………………… 48
M ………………………………………………………… 53
N ………………………………………………………… 61
O ………………………………………………………… 64
P ………………………………………………………… 68
Q ………………………………………………………… 75
R ………………………………………………………… 76
S ………………………………………………………… 81
T ………………………………………………………… 91

U …… 96
V …… 98
W …… 100
Y …… 102
Z …… 103
以数字开头的词条 …… 104

A

A∗ algorithm A星算法

AAL (ambient assisted living) 居家协助

AAV (autonomous aquatic vehicle) 自主水中载具，自主水下航行器

ABA (articulated-body algorithm) 铰接体算法

absolute delay 绝对时延

absolute encoder 绝对(式)编码器

absolute location 绝对位置

absolute rotary encoder 绝对式旋转编码器

absolute velocity 绝对速度

AC (alternating current) 交流

ACA (ant colony algorithm) 蚁群算法

ACC (adaptive cruise control) 自适应巡航控制

acceleration 加速度，加速

acceleration of gravity 重力加速度

acceleration transducer 加速度传感器

accelerometer 加速度计，加速度传感器

accessibility 可达性

accommodation-assimilation 顺应-同化

accommodation limit 适应范围，调节极限

accumulated error 累积误差

accumulative error 累积误差

accuracy 准确度

ACFV (autonomous combat flying vehicle) 自主战斗飞行器

ACO (ant colony optimization) 蚁群优化

acoustic baseline 声学基线

acoustic communication 声学通信

acoustic holography 声全息影像技术

acoustic imaging 声学成像

acoustic model 声学模型

acoustic modem 声频调制解调器

acoustic navigation positioning system 声波导航定位系统

acoustic positioning 声波定位

A

acoustic positioning system 声波定位系统
acoustic sensor 声传感器
acquisition of signal 信号采集，信号获取
acquisition time 采集时间
ACS (ant colony system) 蚁群系统
action 动作
action cycle 动作周期
action description language (ADL) 动作描述语言
action execution 动作执行
action representation 动作表示，动作表征
action selection 动作选择
activation function 激活函数
active guidance 主动制导
active interaction control 主动交互控制
active learning 主动学习
active manipulation for perception 为感知而主动操作，主动感知操作
active sensing 主动感知
active sensor 主动式传感器，有源传感器
active sensor network (ASN) 主动传感器网络
active SLAM (simultaneous localization and mapping) 主动同时定位与建图，主动 SLAM
active steerable wheel 主动可转向轮
active vision 主动视觉
actor-critic method 演员-评论家方法，AC 方法
actuating element 执行元件[单元]，驱动元件[单元]
actuation 驱动，执行，致动
actuation architecture 驱动架构
actuator 驱动器，致动器，执行器
actuator dynamics 驱动器动力学
actuator redundancy 驱动器冗余性
acutance 锐度
AdaBoost 自适应增强算法
adaptability 适应性，适应能力
adaptable 可适应的
adaptable system 自适应系统
adaptation 适应，自适应
adaptive 自适应的
adaptive artificial hand 自适应人工手，自适应假手
adaptive clustering 自适应聚类
adaptive control 自适应控制
adaptive control system 自适应控制系统
adaptive cruise control (ACC) 自适应巡航控制
adaptive expert system 自适应专家系统

adaptive fuzzy logic control 自适应模糊逻辑控制
adaptive grasping 自适应抓取
adaptive identifier 自适应辨识器
adaptive learning 自适应学习
adaptive learning system 自适应学习系统
adaptive measurement 自适应测量
adaptive resonance theory (ART) 自适应共振理论
adaptive search approach 自适应搜索方法
adaptive system 自适应系统
adaptive telemetry system 自适应遥测系统
ADAS (advanced driver assistance system) 先进驾驶员辅助系统
additional load 附加负载
additive manufacturing 增材制造，增量制造，累积制造（3D 打印的另一种说法）
add-one smoothing 加一平滑（算法）
ADF (automatic direction finder) 自动测向器，自动航向检测器，自动探向器
adjacent node 相邻节点
adjoint action 伴随作用
adjoint matrix 伴随矩阵
adjustable pattern generator (APG) 可调模式生成器
ADL (action description language) 动作描述语言
admissible function 容许函数，可取函数
admissible heuristics 容许启发式方法
admittance 导纳
admittance control 导纳控制
admittance matrix 导纳矩阵
ADT (alternating decision tree) 交替决策树
advanced driver assistance system (ADAS) 先进驾驶员辅助系统
advanced manufacturing technology 先进制造技术
adversarial environment 对抗性环境
adversarial search 对抗搜索，敌对搜索
aerial robot 空中机器人
aerial vehicle 飞行器
aerodynamic center 气动中心，空气动力中心
aerodynamic decelerator 气动减速器
aerodynamic force 气动力
aerodynamic stability 气动稳定性
aerodynamics 空气动力学，气体动力学
aeronavigation 领航学，空中导航

aerophare 航空用信标

affine camera 仿射相机

affine coordinate system 仿射坐标系

affine fundamental matrix 仿射基本矩阵

affine geometry 仿射几何

affine group 仿射群

affine layer 仿射层

affine reconstruction 仿射重建[构]

affine space 仿射空间

affine transformation 仿射变换

affixing frame 附着坐标系

affordance 可供性

affordance detection 可供性检测

affordance learning 可供性学习

AFV (autonomous flying vehicle) 自主飞行器

agent 智能体,代理,主体

aggregation 聚集,聚合;(多智能体的)聚集行为

AGI (artificial general intelligence) 通用人工智能,强人工智能

agile manufacturing 敏捷制造

agonistic-antagonistic 竞争-对抗的

agricultural automation 农业自动化

agricultural robot 农业机器人

agricultural robotics 农业机器人学[技术]

AGV (automated guided vehicle) 自动导引车

AHS (automated highway system) 自动公路系统

aided tracking 半自动跟踪,辅助跟踪

airborne 机载的,空中的

airborne vehicle 航空器

air brake 空气制动器,气闸

aircraft dynamics 航空器动力学

air damping 空气阻尼

air guideway 气浮导轨

air traffic control 飞行管制,空中交通控制

AIS (automated intelligence system) 自动化智能系统

alarm sensor 报警传感器

algorithm complexity 算法复杂性,算法复杂度

algorithm convergence 算法收敛性

algorithm decomposition 算法分解

algorithmic approach 算法逼近

algorithmic language 算法语言

algorithm of intelligent planning 智能规划算法

aliasing 混叠

alignment pose 调准位姿

allocentric map 非自我中心地图

all-or-none law 全或无定律

allowable initial state 容许初始状态

ALP (automated language processing) 自动语言处理

alternating current (AC) motor 交流电机

alternating current (AC) servomotor 交流伺服电机

alternating current (AC) torque motor 交流力矩电机

alternating decision tree (ADT) 交替决策树

altimeter 测高仪，高度表

AM (amplitude modulation) 调幅

ambient 环境的，周围的，背景的

ambient assisted living (AAL) 居家协助

ambient intelligence 环境智能

amorphous computing 无定形计算

amplitude 振幅

amplitude modulation (AM) 调幅

AMR (autonomous mobile robot) 自主移动机器人

analog 模拟(量、装置、设备、系统等)

analog camera 模拟相机

analog control 模拟控制

analogical inference 类比推理，类比推断

analogical learning 类比学习

analogical problem solving 类比问题求解

analogical problem space 类比问题空间

analogical reasoning 类比推理

analogical simulation 类比仿真

analog model 相似模型，类比模型

analog signal 模拟信号

analog-to-digital converter 模数转换器

analysis 分析，解析

analytic learning 分析学习

android 类人机器人；(常作 Android) 安卓操作系统

angle 角度

angle gripper 角形夹持器

angle transducer 角度传感器

angular acceleration 角加速度

angular displacement 角位移

angular velocity 角速度

angular velocity transducer 角速度传感器

angular velocity vector 角速度向量

animacy 生命度，有生性

animatronics 电子动物学

ANN (artificial neural network) 人工神经网络

annealed particle filtering (APF) 退火粒子滤波

ant colony algorithm (ACA) 蚁群算法

ant colony optimization (ACO) 蚁群优化

A

ant colony system (ACS) 蚁群系统
ant robotics 蚂蚁机器人学[技术]
antenna 天线
anthropomorphic arm 拟人手臂
anthropomorphic contact 拟人接触
anthropomorphic design 拟人设计
anthropomorphic end-effector 拟人末端执行器
anthropomorphic manipulator 拟人机械臂
anthropomorphic robot 拟人机器人
anthropomorphism 拟人论,拟人观
anthropomorphization 拟人化,人格化
antialiasing 抗混叠
antirust coating 防腐蚀层
anytime algorithm 任意时间算法
a prior knowledge 先验知识
APF (annealed particle filtering) 退火粒子滤波
APF (artificial potential field) 人工势场
APG (adjustable pattern generator) 可调模式生成器
appearance-based approach 基于外观的方法,基于表征的方法
applied force 作用力
apprehension 理解能力
apprenticeship learning 学徒式学习
approximate method 近似方法
approximate model 近似模型
AR (augmented reality) 增强现实
architecture (机器人)体系结构,架构,结构
arc welding robot 弧焊机器人
area image sensor 面型[区域型]图像传感器
arm (机器人)手臂
Aronhold-Kennedy theorem 阿朗浩尔特-肯尼迪定理,三心定理
ART (adaptive resonance theory) 自适应共振理论
articulated-body algorithm (ABA) 铰接体算法
articulated robot 关节(型)机器人,铰接式机器人
articulated soft robot 铰接式软体机器人
articulated soft robotics 铰接式软体机器人学[技术]
articulate-type modular robot 铰接型模块化机器人,关节型模块化机器人
artificial constraint 人工约束,虚约束
artificial ear 人工耳
artificial emotion 人工情感
artificial evolution 人工演化
artificial general intelligence (AGI)

通用人工智能，强人工智能
artificial genome 人工基因组
artificial hand 人工手，假手
artificial intelligence 人工智能
artificial intelligence system 人工智能系统
artificial life 人工生命
artificial limb 义肢，假肢
artificial muscle 人工肌肉
artificial neural network (ANN) 人工神经网络
artificial perception 人工感知
artificial potential field (APF) 人工势场
artificial sampling 人工抽样
artificial skin 人工皮肤
ASN (active sensor network) 主动传感器网络
aspect ratio 幅形比，高宽比，纵横比
assembly 装配
assembly line 装配（流水）线
assembly robot 装配机器人
assistive rehabilitation robot 辅助康复机器人
assistive robot 辅助机器人
assistive robotics 辅助机器人学［技术］
asymmetrical grasping 非对称抓取
asymptotic analysis 渐近分析
asymptotic stability 渐近稳定性
asymptotic stabilization 渐近镇定
asynchronous transmission 异步传输
attached frame 附着坐标系
attained pose 实到［实际］位姿
attention control 注意力控制
attention mechanism 注意力机制
attitude 姿态
attitude control 姿态控制
auction-based method 拍卖法
auditory scene analysis 听觉场景分析
augmented Jacobian matrix 增广雅克比矩阵
augmented reality (AR) 增强现实
authenticity check 真实性检验
autoencoder 自编码器
auto-manual system 自动-手动（切换）系统，半自动系统
automated 自动化的，自动的
automated guided vehicle (AGV) 自动导引车
automated highway system (AHS) 自动公路系统
automated intelligence system (AIS) 自动化智能系统
automated language processing (ALP) 自动语言处理

automated motion planning 自动运动规划
automated reasoning 自动推理
automated task planning 自动任务规划
automatic 自动的
automatic collision detection 自动碰撞检测
automatic control engineering 自动控制工程
automatic control system 自动控制系统
automatic control theory 自动控制理论
automatic direction finder (ADF) 自动测向器,自动航向检测器,自动探向器
automatic end-effector exchange system 自动末端执行器更换系统
automatic identification 自动辨识,自动识别
automatic mode 自动模式
automatic operation 自动操作
automatic programming 自动编程
automatic regulatory system 自动调节系统
automatic tracking 自动跟踪
automation 自动化
automaton 自动机
autonomic computing 自主计算
autonomous agent 自主智能体
autonomous aquatic vehicle (AAV) 自主水中载具,自主水下航行器
autonomous car 自主车
autonomous combat flying vehicle (ACFV) 自主战斗飞行器
autonomous control 自主控制
autonomous exploration 自主探索
autonomous flying vehicle (AFV) 自主飞行器
autonomous logistics 自动物流
autonomous mobile robot (AMR) 自主移动机器人
autonomous navigation 自主导航
autonomous operation 自主操作
autonomous robot 自主机器人
autonomous rover 自主漫游机器人
autonomous system 自主系统,自治系统
autonomous underwater manipulator 自主水下机械臂
autonomous underwater vehicle (AUV) 自主水下载具,自主式水下航行器
autonomous vehicle 自主载具,自主车辆,自主飞行器
autonomy 自主性,自治性,自主能力
autopilot 自动驾驶仪
auto-recharging 自动充电

availability ratio 利用率,可利用系数

available signal-to-noise ratio 可用信噪比

average-pooling 平均池化

avionics 航空电子学

axial magnification 轴向放大率

axis 轴,机械轴;坐标轴

axis angle 关节角度;轴线角

axis-angle representation 轴-角表示法

axis direction 轴线方向

axisymmetrical 轴对称的

azimuth 方位;方位角

B

baby schema　婴儿图式
background noise　背景噪声
back-propagation (BP) neural network　反向传播神经网络
backstepping　反步法
backstepping control　反步控制
backtracking　回溯法
bag-of-words model　词袋模型
ball screw　滚珠丝杠
ball-joint manipulator　球关节机械臂
band-eliminator filter　带阻滤波器
bandit problem　老虎机问题
bang-bang control　砰－砰控制，0－1控制，继电控制，继电器式控制，开关控制
Barbalat's lemma　巴巴拉引理
base　基座
base coordinate system　基座坐标系
base frame　基座坐标架，基座坐标系
baseline drift　基线漂移
Bayes analysis　贝叶斯分析
Bayes classifier　贝叶斯分类器
Bayes decision theory　贝叶斯决策理论
Bayes filter　贝叶斯滤波器
Bayesian belief network　贝叶斯信念网络
Bayesian classifier　贝叶斯分类器
Bayesian decision　贝叶斯决策
Bayesian inference　贝叶斯推理，贝叶斯推断
Bayesian learning　贝叶斯学习
Bayesian machine learning　贝叶斯机器学习
Bayesian network　贝叶斯网络
Bayesian probability　贝叶斯概率
Bayesian reasoning　贝叶斯推理
Bayes' rule　贝叶斯规则
Bayes' theorem　贝叶斯定理
BBR (behavior-based robotics)　基于行为的机器人学[技术]
BCI (brain-computer interface)　脑机接口
beacon　信标

BEAM robotics BEAM 机器人学[技术]

beam search 定向搜索

bearing 方位;轴承

behavioral control 行为控制

behavioral robotics 行为机器人学[技术]

behavior-based control 基于行为的控制

behavior-based robotics (BBR) 基于行为的机器人学[技术]

behavior-based task allocation 基于行为的任务分配

behavior coordination 行为协调

behavior fusion 行为融合

behavior mark-up language (BML) 行为标记语言

behavior primitive 行为基元

behavior selection 行为选择

behavior synthesis 行为合成,行为综合

belief 信念,置信度

belief function 信念函数,置信度函数

belief network 信念网络

belief propagation 信念传播

belief space 信念空间

belief update 信念更新,置信度更新

Bellman equation 贝尔曼方程

Bellman principle of optimality 贝尔曼最优性原理

best-first search 最佳优先搜索

bicycle-type mobile robot 自行车型移动机器人

bidirectional recurrent neural network (BRNN) 双向递归神经网络

bidirectional search 双向搜索

big data 大数据

bilateral teleoperation 双边遥操作

bimetallic strip 双金属片,双金属条

binary image 二值图像

binary mask 二元掩模

binocular SLAM 双目同时定位与建图,双目 SLAM

binocular stereopsis 双目立体视觉

binocular vision 双目视觉

binocular vision measurement 双目视觉测量

bin-picking 容器拣货

bio-inspired actuation 生物启发驱动

bio-inspired climbing 生物启发攀爬

bio-inspired navigation 生物启发导航

bio-inspired robotics 生物启发机器人学[技术]

bio-inspired soft robot 生物启发软体机器人

biological intelligence 生物智能

biologically-inspired robotics 生物启发机器人学[技术]
biomechanics 生物力学
biomechatronics 生物机电一体化
biomedical telemetry 生物医学遥测
biometric authentication 生物鉴别，生物特征认证
biometric identifier 生物特征标识符
biometric matching 生物特征匹配
biometric recognition 生物特征识别
biometric sensor 生物特征传感器
bio-microrobotics 生物微机器人学[技术]
biomimesis 生物拟态
biomimetic gait 仿生步态
biomimetic robot 仿生机器人
biomimetic robotics 仿生机器人学[技术]
biomimetics 生物模仿学
BION (bionic neuron) 仿生神经元
bionic neuron (BION) 仿生神经元
bionics 仿生学
biorobotics 仿生机器人学[技术]
biosonar 生物声呐
biped robot 双足机器人
bipedal locomotion 双足移动
bistatic transducer array 双基站传感器阵列
bi-steerable mobile robot 双转向(轮的)移动机器人
BMI (brain-machine interface) 脑机(器)接口
BML (behavior mark-up language) 行为标记语言
Bode diagram 伯德图
BP (back-propagation) 反向传播
boosting 提升(算法)，加速(算法)
brain-computer interface (BCI) 脑(计算)机接口
brain imaging 脑成像
brain-machine interface (BMI) 脑机(器)接口
brake 制动器，闸，刹车
brake dynamometer 制动测力计
braking system 制动系统
branched kinematic chain 分支运动链
breadth-first search 广度优先搜索
brightness 亮度
BRNN (bidirectional recurrent neural network) 双向递归神经网络
brushless DC electric motor 无刷直流电机
bumper 缓冲器，保险杠

C

cable-driven robot 线驱动机器人

cable-suspended parallel mechanism 悬索并联机构

calibration 校准;标定

calibration error 校准误差;标定误差

calibrator 校准器,标定器

camera 摄影机,照相机,相机

camera alignment 相机调准

camera axis 相机轴

camera calibration 相机标定

camera center 相机中心

camera constant 相机常数

camera coordinate system 相机坐标系

camera external parameters 相机外参数

camera frame 相机坐标系

camera geometry 相机几何

camera internal parameters 相机内部参数

camera intrinsic parameters 相机固有参数,相机内部参数

camera matrix 相机矩阵

camera model 相机模型

camera optics 相机光学

camera parameters 相机参数

camera projection matrix 相机投影矩阵

candidate elimination 候选消除

Canny edge detection 坎尼边缘检测

cantilever 悬臂

capacitance sensor 电容传感器

capacitive encoder 电容(式)编码器

capsule robot 胶囊机器人

cardan joint 万向节

cardan universal joint 十字轴式万向节

cardinal plane 主平面,基平面

car-like mobile robot 类车移动机器人

carrying capacity 承载能力

Cartesian 笛卡尔的,直角坐标的

Cartesian component 笛卡尔分量

Cartesian coordinate frame 笛卡

C

尔坐标系

Cartesian (coordinate) robot 笛卡尔(坐标)机器人

Cartesian impedance control 笛卡尔阻抗控制

Cartesian manipulator 笛卡尔机械臂

Cartesian space 笛卡尔空间

Cartesian stiffness control 笛卡尔刚度控制

CAS (complex adaptive system) 复杂适应系统

case-based reasoning (CBR) 案例式推理,实例推理

caster wheel 万向脚轮,活动脚轮

catadioptric camera 折反射相机

CBR (case-based reasoning) 案例式推理,实例推理

CCD (charge-coupled device) 电荷耦合器件

cellular automata 细胞自动机,元胞自动机(复数形式)

cellular automaton 细胞自动机,元胞自动机

center drive 中心传动

center of action 作用中心

center of equilibrium 平衡中心

center of gravity (COG) 重心

center of gyration 回转中心

center of mass (COM) 质心

center of moment 力矩中心

center of rotation 旋转中心

centralized 集中式的

central pattern generator (CPG) 中枢模式发生器

central projection 中心投影

centric grasping 对中抓取

centrifugal force 离心力

centripetal force 向心力

centroid 几何中心,形心

certain decision 确定性决策

certain reasoning 确定性推理

challenge and reply 询问和应答

chaos theory 混沌理论

chaotic neural network 混沌神经网络

character recognition 字符识别

characteristic frequency 特征频率

characteristic variable 特征变量

Chasles' theorem 沙勒定理

chassis 底盘

chemical vapor deposition (CVD) 化学气相沉积

chrominance 色度

chronometer 精密计时计(经线仪)

circular arc interpolation 圆弧插补

circular motio 圆周运动

circular velocity 圆周速度

circumferential force 圆周力,切

向力
clamping force 夹持力，夹紧力
classical control system theory 经典控制系统理论
classical control theory 经典控制理论
classical mechanics 经典力学
classification （模式）分类
classificator 分级器，分类器
classifier 分类器
cleaning robot 清洁机器人
climb factor 爬升系数
climbing robot 攀爬机器人
close coupling 紧耦合
closed-chain 闭链
closed-form solution 封闭形式的解，封闭解
closed kinematic chain 闭式运动链
closed-loop control 闭环控制
closed-loop gain 闭环增益
closed-loop mechanism 闭环机构，闭链机构
closed-loop robot 闭链机器人
closed-loop system 闭环系统
closed system 封闭系统
closed world assumption 封闭世界假设
cloud computing 云计算
cloud control system 云控制系统
cloud decision-making 云决策
cloud diagnosis 云诊断
cloud learning 云学习
cloud platform 云（计算）平台
cloud robotics 云机器人学[技术]
clustering 聚类；集群
CMM (coordinate measuring machine) 坐标测量机
CMOS (complementary metal oxide semiconductor) 互补金属氧化物半导体
CNC (computer numerical control) 计算机数字控制
CNN (convolution neural network) 卷积神经网络
coalition 联盟
coalition formation 联盟形成，结盟
cobot 协作机器人
coefficient of friction 摩擦系数
coefficient of viscosity 黏性系数
coevolution 协同进化（算法）
COG (center of gravity) 重心
cognition 认知
cognitive control 认知控制
cognitive development 认知发展
cognitive model 认知模型
cognitive robotics 认知机器人学[技术]
cold-light source 冷光源
colidar 激光雷达
collaborative operation 协作操作

collaborative robot 协作机器人

collaborative robotics 协作机器人学[技术]

collaborative workspace 协作工作空间

collective behavior 集体行为

collective intelligence 集体智能

collective robotics 集体机器人学[技术]

collimator 准直仪

collision 碰撞

collision avoidance 避碰

collision detection 碰撞检测

collision-free admissible path 无碰撞可行路径

collision theory 碰撞理论

color camera 彩色相机

color channel 颜色通道

color feature 颜色特征

color fidelity 颜色保真度

color image 彩色图像

color index 颜色索引

color model 色彩模型

color space 色彩空间

color spectrum 色谱

COM (center of mass) 质心

combined guidance 复合制导

command guidance 指令制导

communication protocol 通信协议

companion robot 陪伴机器人

compensability 可补偿性

compensating control 补偿控制

compensating feedback 补偿反馈

compensating signal 补偿信号

competition scenario 竞争场景

complementarity problem 互补性问题

complementary metal oxide semiconductor (CMOS) 互补金属氧化物半导体

complete induction 完全归纳法

completeness 完备性,完整性,完全性

complex adaptive system (CAS) 复杂适应系统

complex control system 复杂控制系统

complex sensor 复合传感器

complexity 复杂性;复杂度

complexity of algorithm 算法复杂度

compliance 柔顺性,柔顺,柔顺度,顺应度

compliance control 柔顺控制

compliance matrix 柔顺矩阵

compliant actuator 柔顺驱动器

compliant assembly 柔顺装配

compliant behavior 柔顺行为

compliant contact model 柔顺接触模型

compliant material 柔性材料
compliant model 柔顺模型
composite decision process 综合决策过程
composite material 复合材料
compound epicyclic reduction gear 两级行星减速器
computational complexity 计算复杂性;计算复杂度
computational efficiency 计算效率
computational error 计算误差
computational intelligence 计算智能
computational learning theory 计算学习理论
computational linguistics 计算语言学
computational model 计算模型
computational neuroethology 计算神经行为学
computational neuroscience 计算神经科学
computed-torque control 计算力矩控制
computer-aided tomography 计算机辅助成像
computer-integrated surgery 计算机集成手术
computer numerical control (CNC) 计算机数字控制
computer vision (CV) 计算机视觉
concave programing 凹形规划法
condition-action rule 条件-动作规则
conditional entropy 条件熵
conditional Gaussian distribution 条件高斯分布
conditional random field (CRF) 条件随机场
conditioner 调节器
condition monitoring 状态监测
condition of balance 平衡条件
configuration 构型,位形;配置
configuration space (C-space) 位形空间,构型空间
conformant planning 一致性规划
conformity error 一致性误差
conjugate gradient method 共轭梯度法
conjugate image 共轭像
consensus 一致性,趋同
conservation of angular momentum 角动量守恒
conservation of energy 能量守恒
conservation of momentum 动量守恒
conservative force 保守力,守恒力
consistent estimator 一致估计量,一致推算子
constellation model 星座模型
constrained equation 约束方程

constrained Lagrangian equation 约束拉格朗日方程
constrained motion 约束运动
constrained optimization 约束优化
constraint 约束
constraint force 约束力
constraint frame 约束坐标系
constraint satisfaction problem (CSP) 约束满足问题
constraint set 约束集合
constraint space 约束空间
construction automation 建筑自动化
construction robot 建筑机器人
constructive induction 构造性归纳
constructive solid geometry (CSG) 构造实体几何,体素构造表示法
contact array 接触阵列
contact deformation 接触形变
contact force 接触力
contact modeling 接触建模
contact sensor 接触传感器
contact stiffenss 接触刚度
contingency 偶然性
contingency planning 应急规划
continuous hidden Markov model 连续隐马尔可夫模型
continuous path control 连续路径控制
continuous rating 连续运转额定值,连续定额
continuous time system 连续时间系统
continuous time-varying feedback 连续时变反馈
continuous transmission frequency modu-lated (CTFM) sonar 连续传输调频声呐
continuous variable dynamic system (CVDS) 连续变量动态系统
continuum 连续统一体,连续体
continuum kinematics 连续体运动学
continuum manipulator 连续体机械臂
continuum medical robot 连续体医疗机器人
continuum robot 连续体机器人
continuum robotics 连续体机器人学[技术]
control 控制
control algorithm 控制算法
control board 控制板
control command 控制命令
control coupling 控制耦合
control error 控制误差
control input 控制输入
controllability 能控性,可控性
control language 控制语言
control law 控制律
controlled variable 被控变量

control loop　控制回路
control mechanism　控制机制
control model　控制模型
control object　控制对象
control objective　控制目标
control performance index　控制性能指标
control program　控制程序
control signaling　控制信令
control software　控制软件
control standard　控制标准
control station　控制站
control system　控制系统
control theory　控制理论
control unit　控制单元
control variable　控制变量
control with fixed set-point　定值控制
convergence　汇聚，收敛
convergence speed　收敛速度
convertor　变流器
convex hull　凸包
convex optimization　凸优化
convex polyhedron　凸多面体
convex programming　凸规划
convex set　凸集
convolution neural network (CNN)　卷积神经网络
cooperation　合作，协作
cooperative manipulation　合作操作
cooperative manipulator　合作机械臂
coordinate　坐标
coordinate descent　坐标下降法
coordinate measuring machine (CMM)　坐标测量机
coordinate system　坐标系
coordinate transformation　坐标变换
coordinate transformation matrix　坐标变换矩阵
coordinated motion　协调运动
coordinated universal time (UTC)　协调世界时，世界标准时间
coordination task　协调任务
coplanarity　共面
Coriolis acceleration　科里奥利加速度，科氏加速度
Coriolis force　科里奥利力，科氏力
corner　角点
corner detection　角点检测
correction method of control system　控制系统校正方法
corrective control　校正控制
corrector system　校正系统
correlation　相关
corresponding pixel　对应像素
cost function　代价函数
cost map　代价地图
Coulomb friction　库仑摩擦
Coulomb law of friction　库仑摩擦定律

C

Coulomb's law 库仑定律
counter 计数器
coupled mode 耦合模式
coupled vibration suppression control 耦合振动抑制控制
coupling 耦合
coupling direction 耦合方向
coupling efficiency 耦合效率
coupling network 耦合网络
covariance 协方差
covariance matrix 协方差矩阵
coverage control 覆盖控制
coverage problem 覆盖问题
CPG (central pattern generator) 中枢模式发生器
Cramer-Rao lower bound 克莱默-拉奥下界
crawling robot 爬行机器人
CRF (conditional random field) 条件随机场
critical damping 临界阻尼
critical path 关键路径
cross-modal learning 跨模态学习
cross-sensitivity 交叉灵敏度
CSG (constructive solid geometry) 构造实体几何,体素构造表示法
CSP (constraint satisfaction problem) 约束满足问题
CTFM (continuous transmission frequency modulated) 连续传输调频
cumulative error 累积误差
cumulative reliability 累积可靠性
current amplifier 电流放大器
current changing ratio 电流变换率
curve fitting method 曲线拟合法
curve motion 曲线运动
curve of error 误差曲线
CV (computer vision) 计算机视觉
CVD (chemical vapor deposition) 化学气相沉积
CVDS (continuous variable dynamic system) 连续变量动态系统
cybercar 智能车
cybernetic system 赛博系统,控制系统
cybernetics 控制论
cyborg 半机械人,赛博格
cycle 周期
cycle time 循环时间
cyclic-balance function 循环平衡功能
cyclopean image 单眼图
cyclopean separation 左右视点分离
cylindrical joint 圆柱关节,圆柱副
cylindrical projection 圆柱法投影
cylindrical robot 圆柱坐标(型)机器人

D

DAI (distributed artificial intelligence) 分布式人工智能

damped natural frequency 阻尼振动固有频率

damper 阻尼器

damping coefficient 阻尼系数

damping control 阻尼控制

damping ratio 阻尼比

dashpot 阻尼器,减震器,缓冲器

data analysis 数据分析

data association 数据关联

data carrier 数据载体

data channel 数据通道

data cleansing 数据清洗

data collection system 数据采集系统

data communication 数据通信

data communication protocol 数据通信协议

data fusion 数据融合

data manipulation language 数据操作语言

data mining 数据挖掘

data modality 数据模式

data packet 数据包

data processing algorithm 数据处理算法

data processing system (DPS) 数据处理系统

data transfer rate 数据传送速率

DBN (deep belief network) 深度信念网络

DBN (dynamic Bayesian network) 动态贝叶斯网络

DC (direct current) 直流

DDD (direct digital drive) 直接数字驱动

DDP (differential dynamic programming) 微分动态规划

dead band 死区

dead reckoning 航位推算[推测]法

dead time 停滞时间,沉寂时间

deadlock 死锁

dead-man switch 失知制动装置

decentering 去中心化

D

decentralized control 分散控制
decision analysis 决策分析
decision boundary 决策边界
decision network 决策网络
decision stump 单层决策树
decision theory 决策论
decision tree 决策树
deconvolution 反卷积
decoupled approach 解耦方法
decoupled path planning 解耦路径规划
decoupled system 解耦系统
decoupling control 解耦控制
dedicated short-range communication (DSRC) 专用短程通信
DEDS (discrete event dynamic system) 离散事件动态系统
deduction theorem 演绎定理
deductive database 演绎数据库
deductive reckoning 演绎推测法
deep belief network (DBN) 深度信念网络
deep learning 深度学习
deep neural network 深度神经网络
deep reinforcement learning 深度强化[增强]学习
deep residual network 深度残差网络
deep transfer learning 深度迁移学习
deflection 挠度
deformable terrain 可变形地形
degree of freedom (DOF) 自由度
degree of maneuverability 可操作度
degree of mobility 移动度,活动度
delay time 延迟时间
deliberative control 慎思控制
delta robot 德尔塔(并联)机器人
Dempster-Shafer theory D-S证据理论
Denavit-Hartenberg method 德纳维-哈登伯格方法,DH方法
Denavit-Hartenberg parameters 德纳维-哈登伯格参数,DH参数
dependability 可靠性
depth 深度
depth-first search 深度优先搜索
depth-first spanning tree 深度优先生成树
depth information 深度信息
depth map 深度图
depth of field 景深
depth reconstruction 深度重构
depth sensor 深度传感器
derivative control 微分控制
detecting device 检测设备
detection resolution 检测分辨率
developmental robotics 发育机器人学[技术]

dexterity 灵巧性

dexterity index 灵巧性指标

dexterous manipulation 灵巧操作

dexterous workspace 灵巧工作空间

DGPS (differential global positioning system) 差分全球定位系统,差分 GPS

DH parameters DH 参数

diagnostic expert system 诊断专家系统

dichotomizing search 二分法搜索

dichotomy 二分法(划分),对分

dielectric elastomer 介电弹性体

difference of Gaussian (DOG) 高斯差分

differential constraint 微分约束

differential drive 差分驱动

differential dynamic programming (DDP) 微分动态规划

differential game 微分对策,微分博弈

differential global positioning system (DGPS) 差分全球定位系统,差分 GPS

differential GPS (DGPS) 差分全球定位系统,差分 GPS

differential kinematics 微分运动学

differentially-driven robot 差分式驱动机器人

digital camera 数字相机

digital control system 数字控制系统

digital image 数字图像

digital image processing 数字图像处理

digital input-output (DIO) 数字输入输出

digital map 数字地图

digital picture 数字图像

digital signal processing (DSP) 数字信号处理

digital simulation 数字仿真

digital transducer 数字传感器

Dijkstra's algorithm 迪杰斯特拉算法

dilation 扩张,膨胀

dilution of precision (DOP) 精度因子

DIO (digital input-output) 数字输入输出

direct current (DC) brush motor 直流有刷电机

direct current (DC) brushless motor 直流无刷电机

direct current (DC) motor 直流电机

direct differential kinematics 正微分运动学

direct-drive (DD) 直接驱动

direct-drive actuation 直接驱动运转

direct-drive (DD) robot 直接驱动

D

机器人

direct force control 直接力控制

direct kinematics 正运动学

disability robot 助残机器人

disaster robot 救灾机器人

disaster robotics 救灾机器人学[技术]

discrete event 离散事件

discrete event dynamic system (DEDS) 离散事件动态系统

discrete Fourier transformation 离散傅里叶变换

discrete programming 离散规划

discrete system 离散系统

discrete-time control system 离散时间控制系统

discrete-time system 离散时间系统

discriminative model 判别模型

disparity 视差

disparity gradient 视差梯度

dispatching 调度,派遣

displacement 位移

display system 显示系统

dissipative structure theory 耗散结构理论

dissipativity theory 耗散理论

distortion 失真;畸变

distortion model 畸变模型

distributed artificial intelligence (DAI) 分布式人工智能

distributed computation 分布式计算

distributed computing 分布式计算

distributed control 分布式控制

distributed control system 分布式控制系统

distributed estimation 分布式估计

distributed localization 分布式定位

distributed network 分布式网络

distributed optimization 分布式优化

distributed parameter system 分布式参数系统

distributed robot system 分布式机器人系统

disturbance 扰动,干扰

disturbed signal ratio 干扰信号比

DOF (degree of freedom) 自由度

DOG (difference of Gaussian) 高斯差分

domestic cleaning robot 家用清洁机器人

domestic robot 家用机器人

domestic robotics 家用机器人学[技术]

DOP (dilution of precision) 精度因子

Doppler radar 多普勒雷达

Doppler sonar 多普勒声呐

Doppler velocity log (DVL) 多普勒测速仪,多普勒计程仪

DPS (data processing system) 数据

处理系统
drive belt 传动带
driver 驱动器;驱动程序;司机
driver assistance system 驾驶员辅助系统
driver monitoring 驾驶员监测
driving flange 传动法兰
driving torque 驱动力矩
drone 无人机
DSC (dynamic sequential control) 动态顺序控制
DSP (digital signal processing) 数字信号处理
DSRC (dedicated short-range communication) 专用短程通信
dual-arm manipulation 双臂操作
dual-arm robot 双臂机器人
duplex feedback 双重反馈
durability 耐久度
durability test 疲劳试验
duty ratio 占空比
DVL (Doppler velocity log) 多普勒测速仪,多普勒计程仪
DWA (dynamic window approach) 动态窗口法
dynamical system 动力系统
dynamic backtracking 动态回溯
dynamic Bayesian network (DBN) 动态贝叶斯网络
dynamic constraint 动态约束
dynamic control 动力学控制
dynamic damper 动力阻尼器
dynamic emulation 动力学仿真
dynamic environment 动态环境
dynamic error 动态误差
dynamic gait 动步态
dynamic look-and-move system 动态视动系统
dynamic model 动力学模型
dynamic programming 动态规划
dynamic range 动态范围
dynamic regulator 动态调节器
dynamic resonance 动态谐振
dynamic response 动态响应
dynamics 动力学
dynamic sequential control (DSC) 动态顺序控制
dynamics of multi-body system 多体系统动力学
dynamic stability 动态稳定性
dynamic system 动态系统
dynamic tactile sensing 动态触觉感知
dynamic thresholding 动态阈值
dynamic window approach (DWA) 动态窗口法

E

EAP (electroactive polymer) 电活性聚合物

earth-centered earth-fixed (ECEF) 地固地心(直角坐标系)

EBL (explanation-based learning) 基于解释的学习

ECEF (earth-centered earth-fixed) 地固地心直角坐标系

echo 回波,回声

echogram 回声深度记录

echolocation 回声定位

echometer 回声探测仪

echo ranging 回波测距

edge detection 边缘检测

educational robot 教育机器人

educational robotics 教育机器人学[技术]

EEG (electroencephalography) 脑电图

effective inertia 有效惯量

effector 执行器

ego-motion 自运动

EHSV (electrohydraulic servo valve) 电动液压伺服阀

EKF (extended Kalman filter) 扩展卡尔曼滤波器

elastic element 弹性元件

elastic impact 弹性碰撞

elasticity 弹性

elastic joint 弹性关节,挠性联轴节

elastic modulus 弹性模量

elastodynamic 弹性动力学的,动弹性

elastomer 弹性体

elastostatic 弹性静力学的

elbow joint 肘关节

elbow manipulator 肘型机械臂

electrical control 电气控制

electrical impedance 电阻抗

electrical suspended gyroscope 静电支承陀螺仪,电悬式陀螺仪

electrical torque 电磁力矩

electric motor 电动马达,电动机

electric servo motor 电动伺服马达,伺服电机

electroactive polymer (EAP) 电活性聚合物

electrochemical actuator 电化学驱动器，电化学致动器

electroencephalogram (EEG) control 脑电控制

electroencephalography (EEG) 脑电图

electrohydraulic servo valve (EHSV) 电动液压伺服阀

electromagnetic actuator 电磁驱动器

electromagnetic compatibility (EMC) 电磁兼容性

electromagnetic compatibility level 电磁兼容性级别

electromagnetic field sensor 电磁场传感器

electromagnetic interference (EMI) 电磁干扰

electromagnetic susceptibility 电磁敏感度

electromechanics 机电学

electromotor 电机

electromyography (EMG) 肌电描记术，肌电图学

electronic distance measuring (EDM) device 电子测距设备

electrorheological fluid 电流变体，电流变液，ER 液

electrotactile 电触觉

elevating screw 上下移动丝杠，升降丝杠

EM (expectation maximization) 期望最大化

embedded control system 嵌入式控制系统

embedded controller 嵌入式控制器

embedded system 嵌入式系统

embodied agent 具身化智能体

embodied cognition 具身化认知

embodiment 具身性

EMC (electromagnetic compatibility) 电磁兼容性

emergence 涌现

emergency braking 紧急制动

emergent behavior 涌现行为

EMG (electromyography) 肌电描记术，肌电图学

EMI (electromagnetic interference) 电磁干扰

emotional robot 情感机器人

emotion-based interaction 基于情感的交互

empirical distribution 经验分布

emulator 仿真器，仿真程序

encoder 编码器

encoding 编码

end-effector 末端执行器

end of arm (EOA) 手臂末端

endoluminal robot 腔内机器人

end-point closed-loop 末端点闭环

end-point open-loop 末端点开环
end-to-end learning 端到端学习
energy stability margin (ESM) 能量稳定裕度
entertainment robot 娱乐机器人
entertainment robotics 娱乐机器人学[技术]
environment map 环境地图
environment model 环境模型
environment modeling 环境建模
environment sensing 环境感知
EOA (end of arm) 手臂末端
epipolar constraint 对极约束,极线约束
epipolar geometry 对极几何
epipolar line 极线,核线
epipolar point 极点,核点
episodic memory 情景记忆
equation of motion 运动方程
equiarm 等臂的
equiaxial 等轴的
equilibrium point 平衡点
equivalent angle-axis representation 等效角轴坐标系表示法
ERF (magnetic rheological fluid) 磁流变体
ergonomics 工效学,人类工程学
error detection 误差检测
error model 误差模型
error propagation 误差传播
error space 误差空间
ESM (energy stability margin) 能量稳定裕度
essential matrix 本质矩阵
estimation error 估计误差
estimation state 估计状态
estimation theory 估计理论
etching 蚀刻
Euclidean distance 欧氏距离
Euclidean geometry 欧氏几何
Euclidean space 欧氏空间
Euler angle 欧拉角
Euler parameters 欧拉参数
Euler's equation 欧拉方程
EVA (extravehicular activity) 舱外活动
evolution 演化,进化
evolutionary algorithm 进化算法
exhaustive algorithm 穷举算法
exhaustive search 穷举搜索
exoskeletal robot 外骨骼机器人
exoskeleton technology 外骨骼技术
exoskeleton-based therapy robot 基于外骨骼的治疗机器人
expectation maximization (EM) algorithm 期望最大化算法
expert consultation system 专家咨询系统
expert control system 专家控制系统

expert system 专家系统

explanation-based generalization 基于解释的泛化

explanation-based learning (EBL) 基于解释的学习

exploding gradient problem 梯度爆炸问题

exploration 探索

exponential decay model 指数衰减模型

exponential distribution 指数分布

exponential growth model 指数增长模型

exponentially stable 指数稳定

exponential smoothing 指数平滑法

extended Kalman filter (EKF) 扩展卡尔曼滤波器

extended Kalman filter (EKF) localization 扩展卡尔曼滤波定位

external force sensing 外力感知

external grasp 外抓取

external guidance 外部制导

external perturbation 外部摄动

exteroceptive sensor 外感受传感器,外传感器

extravehicular activity (EVA) 舱外活动

eye point distance 视点距离

eye-in-hand camera 手眼相机,眼在手上的相机

eye-in-hand visual servoing 手眼视觉伺服,眼在手上的视觉伺服

eye-to-hand camera 固定相机,眼在手外的相机

eye-to-hand visual servoing 固定相机,眼在手外的视觉伺服

F

face recognition 面部识别
facial expression 面部表情
failure control 故障控制
failure detection 故障检测
failure prediction 故障预测
failure rate 故障率
false action 误动作
false negative 假阴性
false positive 假阳性
FastSLAM 快速同时定位与建图，快速 SLAM
fault tolerant control 容错控制
FDM (fused deposition modeling) 熔融沉积成形技术
feasible direction method 可行方向法
feature coding 特征编码
feature detection 特征检测
feature frame 特征帧
feature Jacobian 特征雅克比(矩阵)
feature matching 特征匹配
feature selection 特征选择
feature space 特征空间
feedback 反馈
feedback control 反馈控制
feedback control system 反馈控制系统
feedback factor 反馈系数
feedback image 反馈图像
feedback linearization 反馈线性化
feedback loop 反馈回路
feedback motion planning 反馈运动规划
feedback ratio 反馈比
feedback regulator 反馈调节器
feedback system 反馈系统
feedforward compensation 前馈补偿
feedforward control 前馈控制
feedforward network 前馈网络
FEM (finite element method) 有限元法
ferromagnetism 铁磁性
FES (functional electric stimulation) 功能性电刺激
fiber optic 光纤，光纤维
fiber optic gyroscope (FOG) 光纤

陀螺仪,光纤陀螺
fiber optics 光纤学
field control (磁)场控制
field of view (FOV) 视野
field robot 野外机器人
field robotics 野外机器人学[技术]
filter discrimination 滤波区分能力
find-landmark 地标寻找
fine motion planning 精细运动规划
finger (机器人)手指
fingerprint image 指纹图像
fingertip force sensor 指尖力传感器
fingertip tactile sensor 指尖触觉传感器
finite automaton 有限自动机
finite element method (FEM) 有限元法
finite element model 有限元模型
finite-state automaton 有限状态自动机
finite-state machine (FSM) 有限状态机
fire-fighting robot 灭火机器人,消防机器人
fisheye lens 鱼眼镜头
fitness function 适应度函数
fixed-wing aircraft 固定翼飞机
flapping dynamics 扑翼动力学
flapping flight 扑(翼)飞行
flapping frequency 扑翼频率
flapping robot 扑翼机器人
flapping-wing flight 扑翼飞行
flapping-wing unmanned aerial vehicle (FL-UAV) 扑翼无人机
flash lidar 泛光激光雷达
flexibility 灵活性;柔性,挠性
flexible automation 柔性自动化
flexible gearing 挠性传动装置
flexible joint 柔性关节
flexible link 柔性连杆
flexible manufacturing system (FMS) 柔性制造系统,柔性生产系统
flexible robot 柔性机器人
flight dynamics 飞行动力学
flight simulator 飞行模拟器
FLIR (forward looking infrared) 前视红外仪
flocking 群集
FL-UAV (flapping-wing unmanned aerial vehicle) 扑翼无人机
fluid dynamics 流体动力学
fluxgate sensor 磁通门传感器
fMRI (functional magnetic resonance imaging) 功能性磁共振成像
FMS (flexible manufacturing system) 柔性制造系统,柔性生产系统
FOG (fiber optic gyroscope) 光纤陀螺仪,光纤陀螺
foil gauge 箔式应变计

F

follower 跟随者
following-up system 随动系统
follow-up control 随动控制
foraging behavior 觅食行为
force closure 力封闭
force closure grasp 力封闭抓取
force compliance control 力柔顺控制
force constraint 力约束
force control 力控制
force equilibrium 力平衡
force feedback 力反馈
force feedback control 力反馈控制
force feedback loop 力反馈回路
force manipulability ellipsoid 力可操作性椭球
force reflecting teleoperation 力反射遥操作
force sensing 力感知
force sensing resistor（FSR） 力敏电阻
force sensor 力传感器
force sensor calibration 力传感器标定[校准]
force transducer 力传感器
force vector 力向量
forced vibration 受迫振动
fore arm 前臂
foreground-background segmentation 前景背景分割
forestry automation 林业自动化
forestry robot 林业机器人
forestry robotics 林业机器人学[技术]
formation 编队，队形；形成
formation control 编队控制
forward differential kinematics 正（向）微分运动学
forward dynamics 正向动力学，正动力学
forward instantaneous kinematics 正向瞬时运动学
forward Jacobian matrix 正向雅各比矩阵
forward kinematics 正向运动学，正运动学，运动学正解
forward kinematics model 正向运动学模型，正运动学模型
forward looking infrared（FLIR） 前视红外仪
forward static analysis 正向静力分析
forward statics 正向静力学
four-bar linkage 四连杆结构
Fourier integration 傅里叶积分
Fourier transformation 傅里叶变换
four-legged walking 四腿步行
FOV（field of view） 视野
fractal 分形；分形（机器人）
frame 帧；坐标架，坐标系
frame frequency 帧频

frame of reference 参考坐标系
frame rate 帧频,帧速率
free frequency 固有频率
free gyroscope 自由陀螺仪
free space 自由空间
free vector 自由向量
free vibration 自由振动
Frenet frame 弗勒内坐标系
frequency analysis 频率分析
frequency changer 变频器
frequency converter 频率转换器
frequency divider 分频器
frequency division multiplexing telemetry system 频分多路遥测系统
frequency filter 频率滤波器
frequency ratio 频率比
frequency response 频率响应
frequency response function 频率响应函数
friction coefficient 摩擦系数
friction cone 摩擦锥
friction model 摩擦模型
friction stir welding (FSW) 摩擦搅拌焊接
frictionless point contact 无摩擦点接触
FSM (finite-state machine) 有限状态机
FSR (force sensing resistor) 力敏电阻
FSW (friction stir welding) 摩擦搅拌焊接
Froude number 弗劳德数
functional electric stimulation (FES) 功能性电刺激
functional magnetic resonance imaging (fMRI) 功能性磁共振成像
fundamental matrix 基础矩阵
fused deposition modeling (FDM) 熔融沉积成形技术
fuzzy algorithm 模糊算法
fuzzy clustering 模糊聚类
fuzzy control 模糊控制
fuzzy controller 模糊控制器
fuzzy deduction 模糊演绎
fuzzy inference system 模糊推理系统,模糊推断系统
fuzzy logic 模糊逻辑
fuzzy reasoning 模糊推理
fuzzy theory 模糊理论

F

G

GA (genetic algorithm) 遗传算法

Gabor filter 加博滤波器

gain bandwidth 增益带宽

gain coefficient 增益系数

gain control 增益控制

gait 步态

gait pattern 步态模式

gait recognition 步态识别

gait sensitivity norm 步态灵敏度范数

gait training robot 步态训练机器人

game theory 博弈论

gamma correction 伽马校正

gang control 同轴控制

gantry robot 高架机器人,桁架机器人,龙门机器人

gantry-style robot 高架式机器人,桁架式机器人,龙门式机器人

gas sensor 气敏传感器

gauge factor 应变计灵敏系数

Gaussian kernel 高斯核

Gaussian mixture model (GMM) 高斯混合模型

Gaussian noise 高斯噪声

Gaussian smoothing 高斯平滑

Gauss-Newton method 高斯-牛顿法

Gauss-Newton nonlinear estimation 高斯-牛顿非线性估计

Gauss-Seidel iteration 高斯-赛德尔迭代

GBAS (ground-based augmentation system) 陆基增强系统

GE (genetic evolution) 基因演化

gear drive 齿轮传动

gearing 传动装置

gear ratio 传动比

gear reducer 齿轮减速器

gecko adhesive 壁虎式黏着

general predictive control (GPC) 广义预测控制

generalization 泛化,一般化

generalization ability 泛化能力

generalized coordinate 广义坐标

generalized force 广义力

generalized inertia ellipsoid 广义惯量椭球,广义惯性椭球

generalized inertia matrix 广义惯量矩阵,广义惯性矩阵

generalized Jacobian matrix 广义雅各比矩阵

generalized velocity 广义速度

genetic algorithm (GA) 遗传算法

genetic evolution (GE) 基因演化

genetic programming (GP) 遗传规划

genetic regulatory network (GRN) 基因调控网络

genotype 基因型

geometrical feature 几何特征

geometric path 几何路径

geometric primitive 几何基元

geometric similarity constraint 几何相似性约束

geometrically admissible path 几何容许路径

geometry model 几何模型

gesture recognition 手势识别

giant magnetostrictive actuator (GMA) 超磁致伸缩驱动器

gimbal 万向节

gimbal lock 万向节死锁

global asymptotic stability 全局渐近稳定性

global conditioning index 全局调节指数

global convergence 全局收敛

global localization 全局定位

global motion planning 全局运动规划

global navigation satellite system (GNSS) 全球导航卫星系统

global optimization 全局优化

global planner 全局规划器

global planning 全局规划

global positioning system (GPS) 全球定位系统

global stability 全局稳定性

global task planning 全局任务规划

globally asymptotically stable 全局渐近稳定

globally exponentially stable 全局指数稳定

GMA (giant magnetostrictive actuator) 超磁致伸缩驱动器

GMM (Gaussian mixture model) 高斯混合模型

GNSS (global navigation satellite system) 全球导航卫星系统

GP (genetic programming) 遗传规划

GPC (general predictive control) 广义预测控制

GPS (global positioning system) 全

球定位系统

gradient 梯度

gradient descent 梯度下降法

gradient estimation 梯度估计

gradient estimator 梯度估计器

gradient image 梯度图像

gradient magnitude 梯度幅值

gradient method 梯度法

gradient projection method 梯度投影法

graph-based SLAM 基于图的同时定位与建图,基于图的 SLAM

graphical representation 图形表示

graphical user interface (GUI) 图形用户界面,图形用户接口

graph rigidity 图刚度

graph rigidity theory 图刚度理论

GraphSLAM 基于图的同时定位与建图,图 SLAM

graph theory 图论

grapple fixture 抓牢固定装置

grasp closure 抓取封闭性

grasp mechanics 抓取力学

grasp planning 抓取规划

grasp preshape 抓取预成形

grasp redundancy 抓取冗余

grasp stability 抓取稳定性

grasp synthesis 抓取综合

grasp wrench space 抓取力旋量空间

grasping affordance 抓取可供性

grasping contact model 抓取接触模型

grasp-type gripper 抓握型夹持器

Grassmann geometry 格拉斯曼几何

grating projection 光栅投影

gravity cancellation 重力抵消,重力补偿

gravity compensation 重力补偿

gravity correction 重力修正

gray image 灰度图像

gray level histogram 灰度级直方图

gray scale 灰度

greedy algorithm 贪心算法,贪婪算法

grid 栅格

grid-based map 栅格地图

grid-based model 基于栅格的模型

grid cell 栅格单元

grid filter 栅格滤波器

grid localization 栅格定位

gripper 夹持器

gripping force sensing 夹持力感知

gripping jaw 夹爪

GRN (genetic regulatory network) 基因调控网络

ground-based augmentation system (GBAS) 陆基增强系统

ground control station 地面控制站

ground truth 真值

group synchronization 群同步

guard 防护装置

GUI (graphical user interface) 图形用户界面,图形用户接口

guidance 导引,制导

guidance system 制导系统

guided exploration 导向探索

gyrocompass 旋转罗盘

gyroplane 旋翼机,旋升飞机

gyrorotor 回转体,陀螺转子

gyroscope 陀螺仪

H_2 optimal control H_2最优控制

H_∞ Control H_∞控制

H_∞ optimal control H_∞最优控制

Hall effect 霍尔效应

Hall effect compass 霍尔效应罗盘

Hall effect transducer 霍尔效应传感器

Hall element 霍尔元件

Hamilton-Jacobi-Bellman 哈密顿-雅各比-贝尔曼

Hamilton-Jacobi-Issac 哈密顿-雅各比-艾萨克

Hamilton's principle 哈密顿原理

hand 手

hand-eye calibration 手眼标定[校准]

hands-on cooperative control 手眼协同控制

hand sorting 手动分拣,人工分拣

haptic feedback 触觉反馈

haptic interaction point 触觉交互点

haptic interface 触觉接口

haptic rendering 触觉再现

haptic sensor 触觉传感器

hardness 硬度

harmonic drive 谐波驱动

harmonic drive gear reducer 谐波驱动齿轮减速器

Harris corner 哈里斯角点

Harris detector 哈里斯(角点)检测器

head injury criterion (HIC) 头部伤害指标

head-mounted display 头戴式显示设备

healthcare robot 卫生保健机器人

heat responsive element 热敏元件

helical gearing 螺旋传动

helical joint 螺旋关节,螺旋副

helical motion 螺旋运动

helicopter-type unmanned aerial vehicle (UAV) 直升机型无人机

Hertzian contact 赫兹接触

Hertzian contact model 赫兹接触模型

heterarchical system 异构分层系统

heteroceptive sensor 外感受传感器,外传感器

heterogeneous multi-robot system 异构多机器人系统

heuristic method 启发式方法

heuristic optimization algorithm 启发式优化算法

heuristic search 启发式搜索

HIC (head injury criterion) 头部伤害指标

hidden Markov model (HMM) 隐马尔可夫模型

hidden-state problem 隐藏状态问题

hierarchical agglomerative clustering 层次凝聚聚类

hierarchical control 递阶控制

hierarchical hidden Markov model (HMM) 分层隐马尔可夫模型

hierarchical system 分层系统,递阶系统

hierarchical task network (HTN) 层次[分层]任务网络

high-cut filter 高阻滤波器

high-pass filter 高通滤波器

hill-climbing algorithm 爬山算法

histogram 直方图

histogram of oriented gradients (HOG) 有向梯度直方图

HMM (hidden Markov model) 隐马尔可夫模型

HOG (histogram of oriented gradients) 有向梯度直方图

hologram 全息影像,全息

holonomic constraint 完整性约束

holonomic manipulator 完整机械臂

home automation 家庭自动化

home-based rehabilitation 居家康复

homing action 归航行为

homing guidance 寻的制导,自动引导

homogeneity 同质,同质性;齐次性

homogeneous coordinate 齐次坐标

homogeneous equation 齐次方程

homogeneous image coordinate 齐次图像坐标

homogeneous matrix 齐次矩阵

homogeneous multi-robot system 同质多机器人系统

homogeneous transformation 齐次变换

homogeneous transformation matrix 齐次变换矩阵

homogeneous vector 齐次向量

homography 单应性变换

homography matrix 单应性矩阵

homography projection 单应性投影

Hooke's law 胡克定律

Hooke-type universal joint 胡克万向节

hopping robot 弹跳机器人

Hough transformation 霍夫变换

HTN (hierarchical task network) 层次[分层]任务网络

hue 色调

Huffman tree 霍夫曼树

human arm-like manipulator 类人手臂机械臂

human-centered automation 以人为中心的自动化

human-centered robotics 以人为中心的机器人学[技术]

Hurwitz stable 赫维茨稳定

hybrid dynamic system 混合动态系统

hybrid force control 混合力控制

hybrid force motion control 力/运动混合控制

hybrid force position control 力/位置混合控制

hybrid vision-force control scheme 混合视觉-力控制体系

hybrid visual servoing 混合视觉伺服

hydraulic actuation 液压驱动

hydraulic actuator 液压执行器,液压驱动器

hydraulic cylinder 液压缸

hydraulic drive 液压驱动

hydraulic robot arm 液压机器人臂

hydraulic servo motor 液压伺服马达

hydrodynamics 流体动力学

hypercomplex number 超复数

hyperparameter 超参数

hyper-redundant manipulator 超冗余机械臂

hyper-redundant robot 超冗余度机器人

hysteresis effect 滞后效应

I

IBVS（image-based visual servoing） 基于图像的视觉伺服

ICP（iterative closest point） 迭代最近点算法

identifiability 可辨识性

identification 识别，辨识

IEEE（Institute of Electrical and Electronic Engineers） 电气与电子工程师学会

IEKF（iterated extended Kalman filter） 迭代扩展卡尔曼滤波器

illuminance 照度

illuminance meter 照度计

illuminant 光源

ILP（integer linear programming） 整数线性规划

image 图像

image acquisition 图像采集

image affine coordinate system 图像仿射坐标系统

image analysis 图像分析

image-based visual servoing（IBVS） 基于图像的视觉伺服

image center 图像中心

image contrast 图像对比度

image converting 图像转换

image coordinate 图像坐标

image data 图像数据

image distortion 图像畸变

image feature 图像特征

image fusion 图像融合

image histogram 图像直方图

image matching 图像匹配

image moment 图像矩

image noise 图像噪声

image plane 图像平面

image preprocessing 图像预处理

image processing 图像处理

image processing operator 图像处理算子

image recognition 图像识别

image rectification 图像校正

image region 图像区域

image registration 图像配准

image retrieval 图像检索

image sensor 图像传感器

image sequence 图像序列
image trajectory 图像轨迹
impact dynamics 冲击动力学
impact model 冲击模型
impedance control 阻抗控制
importance sampling 重要性采样
importance weight 重要性权重
impulse generator 脉冲发生器
impulse response 冲激响应
IMS (integrated manufacturing system) 集成制造系统
IMU (inertial measurement unit) 惯性测量单元
incipient failure 初发故障
inclinometer 倾角仪
incremental control strategy 增量控制策略
incremental encoder 增量式编码器
incremental frequency control 增量频率控制
incremental learning 增量式学习
incremental search 增量式搜索
incremental servo-drive 增量式伺服驱动
incremental system 增量系统
independent variable 独立变量
independent-joint control 独立关节控制
indirect control 间接控制
indirect controlled system 间接受控系统
individual axis acceleration 单轴加速度
individual axis velocity 单轴速度
individual joint PID control 单关节比例-积分-微分控制,单关节PID控制
indoor exploration 室内探索
indoor localization system (ILS) 室内定位系统
indoor navigation 室内导航
induction 感应;归纳法,归纳
inductive control 感应控制
inductive encoder 感应式编码器,电感式编码器
industrial control system 工业控制系统
industrial manipulator 工业机械臂
industrial robot 工业机器人
industrial robot cell 工业机器人单元
industrial robot system 工业机器人系
industrial robotics 工业机器人学[技术]
inertia matrix 惯量矩阵,惯性矩阵
inertial guidance 惯性制导

inertial guidance system 惯性制导系统
inertial measurement unit (IMU) 惯性测量单元
inertial navigation 惯性导航
inertial reference frame 惯性参考坐标系
inertial sensor 惯性传感器
inertial surveying system 惯性测量系统
information filter 信息滤波器
information fusion 信息融合
infrared detection system 红外探查系统
infrared guidance control 红外制导控制
infrared image 红外图像
infrared measurement 红外测量
infrared sensor 红外线传感器
in-hand manipulation 手中操作
inner-loop control 内环控制
input-output stability 输入-输出稳定性
input-output-to-state stability (IOSS) 输入-输出-状态稳定性
input-to-state stability 输入-状态稳定性
InSAR (interferometric synthetic aperture radar) 干涉式合成孔径雷达
insect-inspired robot 昆虫启发机器人
insertion sort 插入排序
in-site actuation 在现场驱动
inspector 监测器
installation 安装
instantaneous rotation axis 瞬时转动轴
instantaneous velocity 瞬时速度
Institute of Electrical and Electronic Engineers (IEEE) 电气与电子工程师学会
instrument (自动化)仪表
instrument panel 仪表盘
integer linear programming (ILP) 整数线性规划
integral control 积分控制
integral governing 积分调节
integrated manufacturing system (IMS) 集成制造系统
integrated navigation 组合导航
integrated proximity model 集成近似模型
integrated robotization 集成机器人化
integrated system 集成系统
integrating gyroscope 积分陀螺仪
integration 积分;集成
intellectual science 智能科学
intelligence simulation 智能模拟

intelligent agent 智能自主体，智能体

intelligent automation 智能自动化

intelligent autonomous system 智能自主系统

intelligent backtracking 智能回溯法

intelligent car 智能车

intelligent computer architecture 智能计算机体系结构

intelligent control 智能控制

intelligent control system 智能控制系统

intelligent coordination 智能协调

intelligent learning system 智能学习系统

intelligent machine 智能机器

intelligent peripheral 智能外设

intelligent robot 智能机器人

intelligent system 智能系统

intelligent transportation system (ITS) 智能交通系统

intelligent voice 智能语音

intelligent wheelchair 智能轮椅

intentional cooperation 有意图合作，主动合作

intentionally cooperative system 主动合作系统

interaction 交互

interaction control 交互控制

interaction planning 交互规划

interactive computer system 交互式计算机系统

interactive proof 交互式证明

interactive robot 交互式机器人

interactive system 交互式系统

interface design 交互设计，界面设计

interference 干扰

interferometric fiber optic gyroscope 干涉式光纤陀螺仪

interferometric synthetic aperture radar (InSAR) 干涉式合成孔径雷达

interlock 互锁

internal combustion engine 内燃机

internal force 内力

internal friction 内摩擦

internal grasp 内抓取

internal model principle 内模(型)原理

internal moment 内力矩

internal sensor 内传感器

internal state sensor 内部状态传感器

international space station (ISS) 国际空间站

interoperability 互用性；互操作性

interpolation 插值；插补

interpolation error 插补误差

interpolation point 插补点

intersection collision avoidance 路口避碰

intrinsic error 固有误差

intrinsic image 本征图像

intrinsic noise 固有噪声

intrinsic sensor 本征传感器

intrinsic tactile sensing 本征触觉感知

inverse differential kinematics 逆微分运动学

inverse dynamics 逆动力学

inverse dynamics control 逆动力学控制

inverse Fourier transformation 傅里叶反变换

inverse instantaneous kinematics 逆瞬时运动学

inverse Jacobian 逆雅克比

inverse Jacobian controller 逆雅克比控制器

inverse kinematics 逆运动学

inverse manipulator kinematics 机械臂逆运动学

inverted pendulum system (IPS) 倒立摆系统

IOSS (input-output-to-state stability) 输入-输出-状态稳定性

IPS (inverted pendulum system) 倒立摆系统

IRLS (iteratively reweighted least squares) 迭代重加权最小二乘法

isomorphism 同构

isotropy 各向同性

ISS (international space station) 国际空间站

iterated extended Kalman filter (IEKF) 迭代扩展卡尔曼滤波器

iteration 迭代

iterative closest point (ICP) algorithm 迭代最近邻算法

iterative learning control 迭代学习控制

iterative method 迭代算法

iteratively reweighted least squares (IRLS) 迭代重加权最小二乘法

ITS (intelligent transportation system) 智能交通系统

J

Jacobian 雅克比(矩阵)
Jacobian matrix 雅克比矩阵
Jacobian singular value 雅克比奇异值
Jacobian transpose 雅克比转置
joint 关节
joint angle 关节角
joint axis 关节轴
joint brake 关节制动器
joint elasticity 关节弹性
joint flexibility 关节柔性
joint limit 关节限制
joint rate 关节速率
joint redundancy 关节冗余
joint servo 关节伺服
joint space 关节空间
joint space control 关节空间控制
joint space path 关节空间路径
joint space trajectory 关节空间轨迹
joint torque sensor 关节力矩传感器
joint travel 关节行程
joint travel range 关节行程范围
joint variable 关节变量
jumping robot 跳跃机器人

K

Kalman filter 卡尔曼滤波器
kernel density estimation 核密度估计
kernel function 核函数
kernel method 核方法
kinematic algorithm 运动学算法
kinematic analysis 运动学分析
kinematic calibration 运动学标定［校准］
kinematic chain 运动链
kinematic configuration 运动学构型
kinematic constraint 运动学约束
kinematic control 运动学控制
kinematic coupling 运动学耦合
kinematic equation 运动学方程
kinematic pair 运动副
kinematic redundancy 运动学冗余
kinematics 运动学
kinematic sensor 运动传感器
kinematic simulation 运动学仿真
kinematic singularity 运动学奇异
kinematic skeleton 运动骨架
kinematic synthesis 运动学综合
kinematic tree 运动树
kinesics 人体动作学,体势学
kinesthesia 动觉
kinesthetic sensor 动觉传感器
kinetic control system 动力控制系统
kinetic energy 动能
kinetic model 动力学模型
kinetics 动力学
kinetostatics 运动静力学
kinodynamic planning 运动动力规划
K-means algorithm K 均值算法
K-means clustering K 均值聚类
k-nearest neighbors algorithm (kNN) K 最邻近算法,K 近邻法
kNN (k-nearest neighbors algorithm) K 最邻近算法,K 近邻法
knowledge map 知识地图
knowledge model 知识模型
knowledge representation (KR) 知识表示
knowledge-based vision 基于知识的视觉
KR (knowledge representation) 知识表示

L

ladar　激光雷达

Lagrange formalism　拉格朗日形式论

Lagrange multiplier　拉格朗日乘子

Lagrangian dynamics　拉格朗日动力学

Lagrangian mechanics　拉格朗日力学

Lagrangian operator　拉格朗日算子

landmark　地标

landmark-based map　基于地标的地图

lane changing　车道变更

lane keeping　车道保持

lane tracking　车道跟踪

language model　语言模型

LaSalle's theorem　拉塞尔定理

laser detection　激光探测

laser detection and ranging (ladar)　激光雷达

laser guidance　激光制导

laser imaging radar　激光成像雷达

laser interferometer　激光干涉仪

laser length measure　激光长度测量

laser measurement system　激光测量系统

laser processing robot　激光加工机器人

laser range finder　激光测距仪，激光雷达

laser ranging　激光测距

laser remote sensing　激光遥感

laser scan　激光扫描

laser sensor　激光传感器

laser velocimeter　激光测速仪

laser welding robot　激光焊接机器人

lateral magnification　横向放大率

lattice-type modular robot system　格式模块化机器人系统

layered control architecture　分层控制体系结构

LCD (light coupled device)　光耦合器件

lead screw　导螺杆，丝杠

leader 领航者

learning by demonstration 演示学习

learning control 学习控制

learning control system 学习控制系统

learning controlled robot 学习控制型机器人

learning from history 基于历史学习

learning from human demonstration (LfD) 人类演示学习，人类示范学习

learning robot 学习机器人

least median of squares (LMS) 最小中值二乘法

least squares 最小二乘

least squares estimation (LSE) 最小二乘估计

leg-arm hybrid robot 腿-臂混合机器人

leg exoskeleton 腿外骨骼

legged crawling robot 腿式爬行机器人

legged locomotion 腿式移动

legged robot 腿式机器人

leg-wheel hybrid robot 腿-轮式混合机器人

level of biomimicry 仿生学等级

LfD (learning from human demonstration) 人类演示学习，人类示范学习

librarian robot 图书馆管理机器人

lidar (light detection and ranging) 激光雷达

Lie algebra 李代数

Lie group 李群

life-like robot 类生命机器人，仿生机器人

lifelong localization 终生定位

lifelong mapping 终生地图创建，终生建图

lifelong SLAM 终生同时定位与建图，终生 SLAM

lift-drag ratio 升阻比

light coupled device (LCD) 光耦合器件

light detection and ranging (lidar) 光检测和测距，激光雷达

lighter-than-air unmanned aerial vehicle (LtA－UAV) 轻于空气的无人机

light-section method 光切法

limbed robot 有肢机器人

limit environment 极限环境

limiting device 限位装置

limiting load 极限负载

linear acceleration 线加速度

linear actuator 线性驱动器

L

linear classifier 线性分类器
linear control system 线性控制系统
linear control theory 线性控制理论
linear displacement transducer 直线位移传感器
linear drive 线性驱动
linear interpolation 线性插值
linear inverted pendulum model (LIP) 线性倒立摆模型
linearity 线性度,线性
linearization 线性化
linear matrix inequalities (LMI) 线性矩阵不等式
linear optimization 线性优化
linear quadratic (LQ) optimal control 线性二次型最优控制
linear quadratic Gaussian control (LQG) 线性二次高斯控制
linear system 线性系统
linear velocity 线速度
line displacement 线位移
line drive 线驱动
line extraction 线提取
line of sight (LOS) 视线;瞄准线
line vector 线向量,线矢量
link 连杆
linkage 连杆;链接
link flexibility 连杆柔性
link length 连杆长度
link offset 连杆偏移
link parameter 连杆参数
LIP (linear inverted pendulum model) 线性倒立摆模型
LMI (linear matrix inequalities) 线性矩阵不等式
LMS (least median of squares) 最小中值二乘法
load 负载
load-back method 负载反馈法
load capacity 负载能力
load cell 称重传感器
load force 负载力
loading effect 负载效应
load rate 负载率,载荷率
load ratio 载荷比
load sharing 负载分配
load sharing coefficient 负载分配系数
local asymptotic stability 局部渐近稳定性
local attractivity 局部吸引性
local control 本地控制
local coordinate frame 局部坐标系
localizability 定位能力,可定位性
localizability estimation 定位能力估计,可定位性估计
localizability matrix 定位能力矩

阵，可定位性矩阵

local linearization 局部线性化

locally asymptotically stable 局部渐近稳定

locally degenerate mechanism 局部退化机构

local navigation 局部导航

local optimal 局部最优

local optimization 局部(最)优化

local position control 本地位置控制

local potential function 局部势函数

local reference frame 局部参考坐标系

local regression 局部回归

location set covering problem (LSCP) 位置集合覆盖问题

locator beacon 定位信标

logic-based reasoning 基于逻辑的推理

logic control 逻辑控制

logic sensor system (LSS) 逻辑传感器系统

logistic regression 逻辑回归，逻辑回归分析

logistics robot 物流机器人

logistics automation 物流自动化

long-baseline system 长基线系统

longitudinal velocity 纵向速度

long short-term memory (LSTM) 长短期记忆

long-term SLAM 长期同时定位与建图，长期 SLAM

longwall automation 长壁开采自动化

loop closure 闭环

loosely coupled system 松(散)耦合系统

loosely coupled task 松耦合任务

LOS (line of sight) 视线；瞄准线

loss function 损失函数

lost motion 空转，无效运动

low altitude environment 低空环境

low-earth-orbit operation 近地轨道操作

lower limb wearable system 下肢可穿戴系统

lower pair 低副

lower pair joint 低副关节

low gravity environment 低重力环境

low temperature environment 低温环境

low temperature operation 低温作业

LQ (linear quadratic control) 线性二次最优控制

LQG (linear quadratic Gaussian control) 线性二次高斯控制

LSCP (location set covering problem) 位置集合覆盖问题

LSE (least squares estimation) 最

小二乘估计

LSS (logic sensor system) 逻辑传感系统

LSTM (long short-term memory) 长短期记忆

LtA - UAV (lighter-than-air unmanned aerial vehicle) 轻于空气的无人机

lumped parameter system 集总参数系统

Lyapunov-based control 基于李雅普诺夫的控制

Lyapunov first method 李雅普诺夫第一法,间接法

Lyapunov function 李雅普诺夫函数

Lyapunov method 李雅普诺夫方法

Lyapunov second method 李雅普诺夫第二法,直接法

Lyapunov stable 李雅普诺夫稳定

L

M

machine drive 机器驱动
machine intelligence 机器智能
machine interface 机器接口
machine learning (ML) 机器学习
machine vision 机器视觉
macro-micro manipulator system 宏微机械臂系统
magnetic encoder 磁(性)编码器
magnetic induction 磁感应
magnetic-rheological fluid (ERF) 磁流变液
magnetosensor 磁敏传感器
magnetoencephalography (MEG) 脑磁图
magnetostrictive actuator 磁致伸缩驱动器
Mahalanobis distance 马哈拉诺比斯距离,马氏距离
manipulability 可操作性
manipulability ellipsoid 可操作性椭球
manipulator 机械臂,机械手,操作机,操纵器
manipulator dynamics 机械臂动力学
manipulator kinematics 机械臂运动学
man-machine communication 人机通信
man-machine dialog 人机对话
man-machine interface 人机接口,人机界面
manned vehicle 有人机,有人载具,有人车辆
manual operation 手动操作,人工操作
MAP (maximum a posteriori probability) 最大后验概率
MAPF (multi-agent path finding) 多智能体寻路
map matching 地图匹配
mapping 地图构建,建图,绘制地图
mapping algorithm 建图算法
margin of error 最大容许误差
marginal noise 边缘噪声
marine robot 海洋机器人

markerless motion capture system 无标记运动捕捉系统

market-based method 市场法

Markov assumption 马尔可夫假设

Markov chain 马尔可夫链

Markov chain Monte Carlo (MCMC) 马尔可夫链的蒙特卡罗方法

Markov decision process (MDP) 马尔可夫决策过程

Markov property 马尔可夫特性

Markov random field (MRF) 马尔可夫随机场

Marr-Albus model 马尔-阿布斯模型

Mars exploration rover 火星探测漫游车

Mars Pathfinder 火星探路者

marsupial robot 袋鼠机器人

mass matrix 质量矩阵

mass moment of inertia 质量惯性矩，转动惯量

master side 主端，主机端

master-slave concept 主从概念

master-slave control 主从控制

master-slave manipulator 主从机械臂

master-slave robotic system 主从式机器人系统

match accuracy 匹配准确度

matching error 匹配误差

maximally stable extremal region (MSER) 最大稳定极值区域

maximum a posteriori probability (MAP) estimation 最大后验概率估计

maximum covering location problem (MCLP) 最大覆盖位置问题

maximum likelihood estimation (MLE) 极大似然估计

maximum likelihood estimator (MLE) 极大似然估计器

maximum load 最大负载

maximum payload 最大载荷

maximum principle 极大值原理

maximum tensile force 最大拉力，最大拉伸力

maximum torsional moment 最大扭矩

McKibben artificial muscle 麦吉本式人工肌肉

MCL (Monte Carlo localization) 蒙特卡罗定位

MCLP (maximum covering location problem) 最大覆盖位置问题

MCMC (Markov chain Monte Carlo) 马尔可夫链的蒙特卡罗方法

MDP (Markov decision process) 马尔可夫决策过程

mean-field approximation 平均场近似法

mean-shift 均值漂移
mean-square error (MSE) 均方误差
mean time between failures (MTBF) 平均故障间隔时间,平均无故障时间
mean value 均值
mean value filter 均值滤波器
measurement noise 测量噪声
measuring instrument 检测仪表
Mecanum wheel 麦克纳姆轮
mechanical admittance 机械导纳
mechanical advantage 机械效益
mechanical arm 机械手臂
mechanical governor 机械调速器,机械控制器
mechanical impedance 机械阻抗
mechanical interface 机械接口
mechanical interference 机械干涉
mechanical property 机械性能,力学性能
mechanical resonance 机械共振
mechanical shock 机械冲击
mechanical stiffness 机械刚度
mechanical transmission 机械传动
mechanical wear 机械磨损
mechanical zero 机械零位
mechanically independent finger 机械独立手指,无机械连接手指
mechanically interrelated finger 机械连接手指
mechanism 机构,机械装置;机制
mechanism design 机械设计
mechanism isotropy 机构的各向同性
mechanism optimization design (MOD) 机构优化设计
mechanism synthesis 机构综合
mechanoreceptor 机械性感受器
mechatronics 机电学,机械电子学,机电工程,机电一体化
medical capsule robot 医疗胶囊机器人
medical electrical equipment 医用电气设备
medical electrical system 医用电气系统
medical image 医疗图像,医学图像
medical image processing 医学图像处理
medical image segmentation 医学图像分割
medical robot 医疗机器人
medical robotics 医疗机器人学[技术]
MEG (magnetoencephalography) 脑磁图
MEMS (microelectromechanical system) 微机电系统
mental model 心智模型

M

mesh 网格
M-estimator M 估计器
metal detector 金属探测器
metamorphic anthropomorphic hand 变胞拟人灵巧手
metamorphic hand 变胞(机器人)手
metamorphic mechanism 变胞机构
metamorphic multifingered hand 变胞多指(机器人)手
metastability 亚稳定性
metering equipment 测量仪器
method of approach 渐近法,逐次逼近法
method of approximation 逼近法,近似法
method of least squares (MLE) 最小二乘法
method of linearization 线性化法
method of moments 矩量法
metric camera 量测相机
metric map 尺度地图,度量地图
metric space 度量空间
Metropolis-Hastings algorithm 梅特罗波利斯-黑斯廷斯算法,M-H 算法
microassembly 微装配
microbotics 微机器人学[技术],微尺度机器人学[技术]
microcontroller 微控制器
microelectromechanical system (MEMS) 微机电系统
microfabrication 微制造,微加工,微型品制造
microgravity environment 微重力环境
micromechanical gyroscope 微机械陀螺仪
micromechatronics 微机电工程,微机电一体化
microprocessor 微处理器
microprocessor technology 微处理器技术
microrobotics 微机器人学[技术],微尺度机器人学[技术]
microsurgery 显微手术,显微外科
microsyn 微动同步器,精密自动同步机
microwave radar 微波雷达
microwave remote sensing 微波遥感
midvalue 中值
military robot 军用机器人,军事机器人
millimeter-wave radar 毫米波雷达
milling 铣削
MILP (mixed integer linear programming) 混合整数线性规划
MIMO (multiple-input multiple-output) 多输入多输出
mine clearance robot 地雷排除机

器人

min-entropy 最小熵

minimally invasive surgery 微创外科手术

minimum energy algorithm 最小能量算法

minimum entropy 最小熵

minimum phase system 最小相位系统

minimum spanning tree 最小生成树

mining robot 采矿机器人

mirror neuron 镜像神经元

mixed integer linear programming (MILP) 混合整数线性规划

mixed model 混合模型

ML (machine learning) 机器学习

MLE (method of least squares) 最小二乘法

mobile agent 移动代理

mobile autonomous robot 移动自主机器人

mobile communication network 移动通信网络

mobile computing 移动计算

mobile manipulator 移动机械臂

mobile operating system 移动操作系统

mobile piece-picking robot 移动拣货机器人

mobile platform 移动平台

mobile robot 移动(式)机器人

mobile robotics 移动机器人学[技术]

mobile sensor network (MSN) 移动传感器网络

mobile service robot 移动服务机器人

mobile servicing system (MSS) 移动服务系统

mobile terminal 移动终端

mobile wireless sensor network (MWSN) 移动无线传感器网络

MOD (mechanism optimization design) 机构优化设计

modal analysis 模态分析

modal stiffness 模态刚度

model 模型

model-based adaptive controller 基于模型的自适应控制器

model-free control 无模型的控制

model identification 模型辨识

model learning 模型学习

model predictive control (MPC) 模型预测控制

model reference adaptive control (MRAC) 模型参考自适应控制

model reference adaptive system (MRAS) 模型参考自适应系统

model uncertainty 模型不确定性

modern control theory 现代控制理论
modular design 模块化设计
modular robot 模块化机器人
modular robotics 模块化机器人学[技术]
modularity 模块性(度)
modularization 模块化
modulation 调制
modulation rate 调制速率
module 模件,组件,模块
modulus of elasticity 弹性模量
moment 力矩
moment of inertia 转动惯量,惯性矩,惯性力矩
monocular range finder 单目测距仪
monocular SLAM 单目同时定位与建图,单目 SLAM
monocular vision 单目视觉
monocular vision measurement 单目视觉测量
Monte Carlo filter 蒙特卡罗滤波器
Monte Carlo filtering 蒙特卡罗滤波
Monte Carlo localization (MCL) 蒙特卡罗定位
Monte Carlo method 蒙特卡罗方法
morphological communication 形态学通信
morphological computation 形态学计算
morphological operator 形态算子
morphology 形态学
motion capture 运动捕捉
motion control 运动控制
motion controller 运动控制器
motion detection 运动检测
motion estimation 运动估计
motion field 运动场
motion model 运动模型
motion parallax 运动视差
motion planning 运动规划
motion primitive 运动基元
motion system 运动系统
motion trajectory 运动轨迹
motor 马达;矩向量
motor algebra 矩向量代数
motor evoked potential 运动诱发电位
motor inertia 马达惯量,电机惯量
motor schema 动作图式,运动模式
motor torque 马达转矩,电机转矩
motor torque constant 马达转矩常数,电机转矩常数
movement detection 运动检测
MPA (multi-layer piezoelectric actuator) 多层压电驱动器
MPC (model predictive control) 模型预测控制

MRAC (model reference adaptive control) 模型参考自适应控制
MRF (Markov random field) 马尔可夫随机场
MRS (multi-robot system) 多机器人系统
MRTA (multi-robot task allocation) 多机器人任务分配
MSER (maximally stable extremal region) 最大稳定极值区域
MSS (mobile servicing system) 移动服务系统
MTBF (mean time between failures) 平均故障间隔时间,平均无故障时间
MTSP (multiple traveling salesman problem) 多旅行商问题
multi-agent path finding (MAPF) 多智能体寻路
multi-agent system 多智能体系统
multi-arm robot system 多臂机器人系统
multibody dynamics 多体(系统)动力学
multi-camera system 多相机系统
multi-channel feedback 多路反馈
multidirectional pose accuracy 多方向位姿准确度
multifingered dexterous hand 多指灵巧手
multi-layer piezoelectric actuator (MPA) 多层压电驱动器
multilegged robot 多腿机器人
multi-loop control system 多回路控制系统
multi-loop feedback 多回路反馈
multimodal 多模态的
multimodal interaction 多模态交互
multimodal perception 多模态感知
multiple-input multiple-output (MIMO) 多输入多输出
multiple mobile robots system 多移动机器人系统
multiple traveling salesman problem (MTSP) 多旅行商问题
multiplexer 多路转接器
multi-query planner 多询规划器
multi-robot control 多机器人控制
multi-robot cooperation 多机器人合作
multi-robot coordination 多机器人协调
multi-robot formation 多机器人编队
multi-robot localization 多机器人定位
multi-robot mapping 多机器人建图
multi-robot path planning (MRPP) 多机器人路径规划

multi-robot routing problem 多机器人路由问题

multi-robot scheduling 多机器人调度

multi-robot SLAM 多机器人同时定位与建图,多机器人 SLAM

multi-robot system (MRS) 多机器人系统

multi-robot task allocation (MRTA) 多机器人任务分配

multi-sensor environment modeling 多传感器环境建模

multi-sensor fusion 多传感器融合

multisensory fusion architecture 多传感器融合架构

multitarget observation 多目标观测

multi-variable system 多变量系统

multi-vehicle coordination 多车辆协调

muscle myoelectrical signal 肌肉肌电信号

musculoskeletal walking model 肌肉骨架行走模型

MWSN (mobile wireless sensor network) 移动无线传感器网络

N

nanoelectromechanical system 纳米电子机械系统
nanoengineering 纳米工程
nanoimprint 纳米压印
nanorobotic manipulator 纳米机器人机械臂
nanorobotic system 纳米机器人系统
nanorobotics 纳米机器人学[技术]
nanotechnology 纳米技术
natural frequency 自然频率
natural language processing (NLP) 自然语言处理
navigation 导航
navigation function 导航函数
ND (nearness diagram) 近域图,近似图
nearest-neighbor association 最近邻关联
nearest-neighbor method 最近邻方法
nearness diagram (ND) 近域图,近似图
nearness diagram navigation 近似图导航
network topology 网络拓扑
network topology control 网络拓扑控制
networked communication 网络通信
networked multiple mobile robot system 网络化多移动机器人系统
networked robot system 网络化机器人系统
networked robotics 网络化机器人学[技术]
networked teleoperation 网络遥操作
neural implant 神经植入
neural interface 神经接口
neural-informatics 神经信息学
neural machine translation (NMT) 神经网络机器翻译
neural network (NN) 神经网络
neural processing 神经处理
neural Turing machine 神经图

灵机
neuro-control 神经控制
neuroethology 神经行为学
neuromorphic engineering 神经形态工程学
neuron 神经元
neuroprosthetics 神经义肢技术
neurorobotics 神经机器人学[技术]
Newton-Euler equations 牛顿-欧拉方程
Newton-Raphson method 牛顿-拉夫森方法
Newton's equation 牛顿方程
NLP (natural language processing) 自然语言处理
NMT (neural machine translation) 神经机器翻译
NN (neural network) 神经网络
non-autonomous system 非自治系统,非自主系统
non-contact therapy robot 非接触治疗机器人
non-deterministic polynomial-time (NP) 非确定性多项式时间,NP
non-deterministic polynomial-time complete (NPC) 非确定性多项式时间完全,NPC
non-deterministic polynomial-time hard (NP-hard) 非确定性多项式时间难,NP难
non-deterministic time-delay 非确定性时延
non-grasp-type gripper 非抓握型夹持器
nonholonomic constraint 非完整(性)约束
nonholonomic mobile robot 非完整(约束)移动机器人
nonholonomic motion planning 非完整(约束)运动规划
nonlinear autonomous system 非线性自治系统
nonlinear computational instability 非线性计算不稳定性
nonlinear control system 非线性控制系统
nonlinear dynamic inversion 非线性动态逆(方法)
nonlinear feedback 非线性反馈
nonlinear force function 非线性强制[迫]函数
nonlinear least squares 非线性最小二乘
nonlinear observer 非线性观测器
nonlinear optimal control 非线性最优控制
nonlinear optimization 非线性优化
nonlinear programming 非线性

N

规划

nonlinear system 非线性系统

nonlinear vibration 非线性振动

nonlinearity 非线性

non-maxima suppression 非极大值抑制

non-uniform rational B-spline 非均匀有理B样条

normal operating condition 正常操作条件

normalized energy stability margin (ESM) 归一化能量稳态裕度

NP (non-deterministic polynomial-time) 非确定性多项式时间,NP

NPC (non-deterministic polynomial-time complete) 非确定性多项式时间完全,NPC

numerical analysis 数值分析

numerical control system 数字控制系统

numerical method 数值方法

numerical solution 数值解

nursing robot 护理机器人

Nyquist criterion 奈奎斯特判据

Nyquist diagram 奈奎斯特图

Nyquist stability criterion 奈奎斯特稳定性判据

O

object 目标;物体;对象

object coordinate system 目标坐标系;物体坐标系

object detection 目标检测;物体检测

object frame 目标坐标系;物体坐标系

object geolocation 目标定位;物体定位

object identification 目标标识;物体标识

object learning 目标学习;物体学习

object model 目标模型;物体模型;对象模型

object-oriented programming (OOP) 面向对象的编程

object recognition 目标识别;物体识别

object representation 目标表示;物体表示;对象表示

observability 能观(测)性

observation matrix 观测矩阵

observation model 观测模型

obstacle 障碍,障碍物

obstacle avoidance 避障,障碍物躲避

obstacle avoidance path modification 避障路径修正

obstacle detection 障碍物检测

obstacle potential field 障碍物势场

obstacle region 障碍物区域

occlusion 遮挡

occlusion analysis 遮挡分析

occupancy grid 占据栅格

occupancy grid map 占据栅格地图

occupancy map 占据地图

OCR (optical character recognition) 光学字符识别

octopus robot 章鱼机器人

octree 八叉树

odometer 里程表,里程计

odometry 测程法;里程计

offline calculation 离线计算

offline motion planning 离线运动规划

offline path planning 离线路径规划

offline programming 离线编程

offline task planning 离线任务规划

offline trajectory planning 离线轨迹规划

offset ratio 偏移系数

offsite robotics 现场外机器人学[技术]

omni-bearing navigation system 全方位(角)导航系统

omni-directional camera 全向相机

omni-directional mobile mechanism 全向移动机构

omni-directional mobile robot 全向移动机器人

omni-directional vehicle 全向车辆,全向载具

omnimobile robot 全向移动机器人

one-dimensional space 一维空间

one-dimensional stress 一维应力

one-shot learning 一次性学习

online machine learning 在线机器学习

online motion planning 在线运动规划

online planning 在线规划

online trajectory planning 在线轨迹规划

ontogenetic robotics 个体发育机器人学[技术]

OOP (object-oriented programming) 面向对象的编程

open agent architecture 开放式智能体架构

open kinematic chain 开式运动链

open-loop control 开环控制

open-loop controller 开环控制器

open-loop mechanism 开环机构;开环机制

open roboethics initiative 开放式机器人伦理倡议

open-source robotics (OSR) 开源机器人学[技术]

open system 开放系统

open world assumption 开放世界假设

operating environment 运行环境,操作环境

operating instruction 操作指令

operating point 操作点

operating space 操作空间

operating system (OS) 操作系统

operational reliability 运行可靠性

operational space 操作空间

operational space control (OSC) 操

作空间控制

operational space inertia matrix 操作空间惯性矩阵

operation control unit 操控单元

operator 操作者;算子;操作器

optical character recognition (OCR) 光学字符识别

optical encoder 光(学)编码器

optical flow 光流法

optical flow constraint equation 光流约束方程

optical information processing 光信息处理

optical motion capture system 光学运动捕捉系统

optical pick-up system 光学拾波器系统

optical plummet 光学垂直仪,光学对中器

optical radiation 光辐射

optical sensor 光敏传感器,光学传感器

optimal control 最优控制

optimal control theory 最优控制理论

optimal path 最优路径

optimal planning 最优规划

optimal solution 最优解

optimal task assignment 最优任务分配

optimization (最)优化

optimization method (最)优化方法

orbital robot 轨道机器人

orbital robotics 轨道机器人学[技术]

ordering constraint 顺序约束

orientation 姿态,朝向,方向

orientation accuracy 姿态准确度

orientation angle 姿态角,方向角

orientation control 姿态控制

orientation quaternion 方向四元数

orientation repeatability 姿态重复性

orientation singularity 姿态奇异性

ornithopter 扑翼机

orthesis 矫形器,矫正法

orthogonal group 正交群

orthogonal matrix 正交矩阵

orthogonal rotation matrix 正交旋转矩阵

orthogonal transformation 正交变换

orthographic projection 正交投影

orthonormal matrix 标准正交矩阵

orthopaedic surgical robot 骨科手术机器人

OS (operating system) 操作系统

OSC (operational space control) 操

作空间控制

OSR（open-source robotics） 开源机器人学[技术]

outdoor robotics 户外机器人学[技术]

outlier 异常值,极端值,离群值

outer-loop control 外环控制

overactuated system 过驱动系统

over-approximation 过近似

overconstrained mechanism 过约束机构

overcorrection 过校正,过调

overdamped system 过阻尼系统

overload 过载

oversampling 过采样

overshooting 超调

P

painting robot 喷涂机器人
palletizing robot 码垛机器人
palstance 角速度
panning （移动摄像机）追拍，平移
pan-tilt camera 左右-上下移动相机，云台相机
parallel 并联；并行；平行
parallel algorithm 并行算法
parallel axis theorem 平行轴定理
parallel gripper 平行夹持器
parallel kinematic machine（PKM） 并联运动机床
parallel link robot 并联杆式机器人
parallel manipulator 并联机械臂
parallel mechanism 并联机构
parallel projection 平行投影
parallel robot 并联机器人
parallelism 并行性，平行性
parameter 参数
parameter estimation 参数估计
parameter identification 参数辨识
parametric equation 参数方程
parametric model 参数模型
parametric uncertainty 参数不确定性
parcel handling robot 包裹处理机器人
parking assistance system 泊车辅助系统
partially observable Markov decision process（POMDP） 部分可观测马尔可夫决策过程
particle deprivation problem 粒子匮乏问题
particle filtering（PF） 粒子滤波
particle swarm optimization（PSO） 粒子群优化
passive compliance 被动柔顺性
passive dynamic walking 被动动态行走
passive localization 被动定位
passive sensor 无源传感器
passive SLAM 被动同时定位与建图，被动 SLAM
passive training 被动训练

passive vision 被动视觉
passivity 无源性，被动性
passivity-based control 无源控制
path 路径
path acceleration 路径加速度
path accuracy 路径准确度
path optimization 路径优化
path planning 路径规划
path repeatability 路径重复性
path velocity accuracy 路径速度准确度
path velocity fluctuation 路径速度波动
pattern matching 模式配对，模式匹配
pattern recognition 模式识别
payload 有效负载，有效载荷
PBVS (position-based visual servoing) 基于位置的视觉伺服
PCWF (point contact with friction) 有摩擦点接触
PD control 比例-微分控制
PDOP (positional dilution of precision) 位置精度因子
peg-in-hole task 轴孔任务
pendant 示教盒，示教器
pendulus gyroscope 钟摆式陀螺仪
perception 感知
perception system 感知系统
performance evaluation 性能评价
performance testing 性能测试
period 周期，循环；时期
permanent magnetic DC motor 永磁式直流电机
permeameter 磁导计
persistence of vision 视觉暂留
perspective camera 透视投影相机
perspective projection 透视投影
perspective transformation 透视变换
perturbation 摄动
perturbation analysis 摄动分析
pervasive computing 普适计算
Petri net 佩特里网
phase 相位
phase difference 相位差
phase plane method 相平面法
pheromone 信息素
photo encoder 光电编码器
photoconductive cell 光电导管；光敏电阻
photoconductive element 光敏元件
photodiode 光电（光敏，光控）二极管
photoelectric effect 光电效应
photoelectric sensor 光电传感器
photogrammetry 摄影测量法
photomodulator 光调制器
photo-potentiometer 光电电位器
photoreceptor 光感受器，感光器

photoresistor 光敏电阻器,光电导管

photoswitch 光开关

physical vapor deposition (PVD) 物理气相沉积

pick-and-place operation 拾放操作

picture compression 图像压缩

picture distortion 图像畸变

picture element (pixel) 像素,像元

picture processing 图像处理

picture segmentation 图像分割

PID (proportional-integral-derivative) 比例-积分-微分

PID control law 比例积分微分控制律,PID 控制律

PID gain tuning 比例积分微分增益整定,PID 增益整定

piecewise linearization 分段线性化

piezo crystal 压电晶体

piezo effect 压电效应

piezocoupler 压电耦合器

piezoelectric accelerometer 压电式加速度仪

piezoelectric actuator 压电执行器,压电驱动器

piezoelectric ceramic 压电陶瓷

piezoelectric material 压电材料

piezoelectric motor 压电电机

piezoelectric polymer 压电聚合物

piezoelectric sensor 压电式传感器

piezoelectric stack 压电叠堆

pilot signal 控制(导频、指示)信号

pinch effect 收缩(收聚)效应

pinhole camera 针孔相机

pinhole camera model 针孔相机模型

pipeline robot 管道机器人

pitch 俯仰;高度;音量

pitch angle 俯仰角

pixel 像素

pixel coordinate system 像素坐标系

PKM (parallel kinematic machine) 并联运动机床

planar joint 平面关节,平面副

planar projection 平面投影

planar two-arm system 平面二臂系统

plane homogeneous coordinates 平面齐次坐标

planet gear 行星齿轮

planetary gear reduction 行星齿轮减速

planning 规划

planning algorithm 规划算法

planning under uncertainty 不确定性规划

playback operation 示教再现操

作,重现操作
playback robot 示教再现机器人,重现机器人
PLC (programmable logic controller) 可编程逻辑控制器
Plücker coordinate 普吕克坐标
pneumatic actuator 气动执行器,气动驱动器
pneumatic cylinder 气缸
pneumatic motor 气动马达
pneumatic system 气动系统
point-based value iteration 基于点的值迭代
point cloud 点云
point cloud registration 点云配准
point cloud representation 点云描述,点云表达
point contact with friction (PCWF) 有摩擦点接触
point contact without friction (PwoF) 无摩擦点接触
point estimation 点估计
point feature histogram 点特征直方图
point-to-point (PTP) 点到点
point-to-point (PTP) control 点位控制
polar robot 极坐标(型)机器人
polar system 极坐标系
pole 极点
pole assignment 极点配置
policy iteration 策略迭代
polymeric actuator 聚合物驱动器
polynomial time 多项式时间
polynomial trajectory 多项式轨迹
POMDP (partially observable Markov decision process) 部分可观测马尔可夫决策过程
Pontryagin's maxmum principle 庞特里亚金极大值原理
Pontryagin's minmum principle 庞特里亚金极小值原理
pool cleaning robot 水池清洁机器人
Popov hyperstability 波波夫超稳定性
pose 位姿
pose accuracy 位姿准确度
pose estimation 位姿估计
pose overshoot 位姿超调(量)
pose repeatability 位姿重复性
pose stabilization time 位姿镇定时间
pose-to-pose control 点位控制,位姿到位姿的控制
pose transformation 位姿变换
position 位置
position accuracy 位置准确度
position and orientation 位置与姿态,位姿

position control 位置控制
position encoder 位置编码器
position feedback 位置反馈
position repeatability 位置重复性
position singularity 位置奇异性
position tolerance 位置公差
position transducer 位置传感器[变送器]
position vector 位置向量
positional dilution of precision (PDOP) 位置精度因子
position-based visual servoing (PBVS) 基于位置的视觉伺服
position-force hybrid control 力位混合控制
positioning accuracy 定位准确度
positioning repeatability 定位重复性
positioning system 定位系统
posture 姿态,位姿
potential energy 势能
potential field 势场,位场
potential field method 势场法
potential function 势函数
power system 能源动力系统
powered suit of armor 动力机甲
pragmatics 语用学
precedence constraint 优先约束
precision 精度,精确度
precision lathe 精密车床
predictive control 预测控制
prehensile grasping 适于抓握的抓取
prehensile pattern 适于抓握的模式
prehension 抓持能力
pressure angle 压力角
pressure gauge 压力计
pressure indicator 压力指示器
pressure relay 压力继电器
pressure sensor 压力传感器
Prewitt operator 普里威特算子
primary axis 主关节轴
primary control element 主控元件
principal axis 主轴
principal point 主点
principle of virtual work 虚功原理
prismatic joint 棱柱关节,平移关节,移动副
PRM (probabilistic roadmap method) 概率路线图法
PRN (pseudo-random noise) 伪随机噪声
probabilistic fusion 概率融合
probabilistic method 概率法
probabilistic roadmap method (PRM) 概率路线图法
probabilistic robotics 概率机器人学[技术]
probability hypothesis density 概

率假设密度

procedure-oriented language 面向过程语言

process control system 过程控制系统

product of inertia 惯性积

professional service robot 专业服务机器人

program 程序

program control 程序控制

program-controlled industrial robot 程控工业机器人

programmable assembly machine 可编程装配机器

programmable controller 可编程控制器

programmable logic controller (PLC) 可编程逻辑控制器

programmable measuring apparatus 可编程测量仪器

programmable power supply 可编程电源供应

programmable pulse generator 可编程脉冲发生器

programmable robot 可编程机器人

programmable terminal 可编程终端

programmed automatic test equipment 程控自动检测设备

programmed computer 程控计算机

programmed numerical control 程控数字控制

programmed pose 编程位姿,指令位姿

programming environment 编程环境

program verification 程序验证

projection 投影

projection coordinate system 射影坐标系

projection drawing 投影图

projection geometry 射影几何学

projection matrix 投影矩阵

projection model 投影模型

projection space 射影空间

projection transformation 投影变换,射影变换

proportional action coefficient 比例作用系数

proportional component 比例环节

proportional control 比例控制,P控制

proportional-derivative (PD) control 比例微分控制,PD控制

proportional-integral-derivative (PID) control 比例积分微分控制,PID控制

proportional-integral-derivative (PID) regulator 比例积分微分调节器,PID调节器

proprioception 本体感受，内部感受

proprioceptive sensor 本体感受传感器，内部传感器

propulsion system 推进系统

prosthetic device 假肢设备

prosthetic hand 仿生手，假手

protected area 保护区

protection level 保护等级

protective device 保护装置

protective stop 保护性停止

prototype 样机，模型机，原型

prototype testing 样机实验

proximity sensor 接近传感器

pseudocode 伪代码

pseudo-color 伪彩色

pseudo-correspondence 伪相关，伪对应

pseudo-random noise (PRN) 伪随机噪声

pseudo-range 伪距

PSO (particle swarm optimization) 粒子群优化

PTP (point-to-point) 点到点

publish-subscribe 发布-订阅

pulse-code-modulation telemetry system 脉冲编码调制遥测系统

pulse-counting equipment 脉冲计数设备

pulse engine 脉冲发动机

pulse rate 脉冲频率

pulse width 脉冲宽度

PVD (physical vapor deposition) 物理气相沉积

PwoF (point contact without friction) 无摩擦点接触

Q

QFT（quantitative feedback theory） 定量反馈理论

quadrature encoder 正交编码器

quadrotor 四旋翼飞行器

quadtree 四叉树

quantitative feedback theory（QFT） 定量反馈理论

quaternion 四元数

queueing network（QN） 排队网络

queueing（queuing）theory 排队论

quick forging manipulator 快锻机械臂

R

radar　雷达
radar resolution　雷达分辨率(力)
radial basis function (RBF)　径向基函数
radial basis function (RBF) network　径向基函数网络
radial distortion　径向畸变
radiation environment　辐射环境
radiation source　辐射源
radio control　无线电控制
radio direction finder (RDF)　无线电测向
radio frequency (RF) beacon　无线电信标
radio frequency (RF) identifier tag　射频标识
radio frequency identification (RFID)　射频识别
radio-location equipment　无线电定位设备
radio-navigation　无线电导航
random process　随机过程
random sample consensus (RANSAC)　随机抽样一致性算法
random test　随机试验
random tree　随机树
random variable　随机变量
randomized sampling　随机抽样
range　(值)域;极差,全距;量程
range and bearing sensor　距离和方位传感器
range finder　测距仪
range sensing　距离感知
ranging module　测距模块
ranging sonar　测距声呐
RANSAC (random sample consensus)　随机抽样一致性算法
rapidly-exploring dense tree (RDT)　快速搜索稠密树
rapidly-exploring random tree (RRT)　快速搜索随机树
rated current　额定电流
rated load　额定负载
rated power　额定功率
rated pressure　额定压力
rated speed　额定速率

rated velocity 额定速度
rated voltage 额定电压
rate gyroscope 速率(微分)陀螺仪
rate-integrating gyroscope (RIGS) 速率积分陀螺仪
rate-of-turn gyroscope 角速度(积分)陀螺仪
rate sensor 速率传感器
rating test 标定试验
ratio of damping 阻尼系数
ratio of gear 变速比,传动比
ratio of revolution 转速比
ratio of transmission 传动比
RBF (radial basis function) 径向基函数
RBM (restricted Boltzmann machine) 受限玻尔兹曼机
RCC (remote center compliance) 远中心柔顺性
RCS (real-time control system) 实时控制系统
RDF (radio direction finder) 无线电测向
RDT (rapidly-exploring dense tree) 快速搜索稠密树
reachability 可达性
reachable and dexterous workspace 可达且灵巧的工作空间
reachable workspace 可达工作区
reactive control 反应式控制
reactive robot system 反应式机器人系统
real-time (realtime) 实时
real-time control 实时控制
real-time control system (RCS) 实时控制系统
real-time data requisition 实时数据采集
real-time processing 实时处理
real-time system 实时系统
reconfigurable mechanism 可重构机构
reconfigurable robot 可重构机器人
rectangular robot 直角坐标(型)机器人
rectified linear unit (ReLU) 线性修正单元
recurrent neural network (RNN) 递归神经网络
recursive filtering 递归滤波
recursive least squares filter 递归最小二乘滤波器
recursive Newton-Euler algorithm (RNEA) 迭代牛顿-欧拉算法
reduction-transmission system 减速传动系统
redundancy 冗余,冗余度
redundancy rate 冗余率
redundant degree of freedom (DOF)

冗余自由度

reference frame 参考坐标系

reference plane 参考面,基准面

reference point 参考点

reflective beacon 反射式信标

region growth procedure 区域生长过程

region of interest (ROI) 感兴趣区域

register ratio 机械传动比

regulation control 调节控制,整定控制

rehabilitation robot 康复机器人

rehabilitation training 康复训练

reinforcement learning (RL) 强化学习

relative motion 相对运动

relative movement 相对运动

relative orientation 相对朝向,相对姿态

relaxation algorithm 松弛算法

reliability 可靠性

ReLU (rectified linear unit) 线性修正单元

remote center compliance (RCC) 远中心柔顺性

remote control 远程控制,遥控

remote control system 遥控系统

remote handling 遥处理

remotely operated tool (ROT) 遥控操作工具

remotely operated vehicle (ROV) 遥控无人潜水器,水下机器人

remotely piloted vehicle (RPV) 遥控飞行器

remote manipulator 遥控机械臂

remote metering system 遥测系统

remote operation 远程操作

remote sensing 遥感

remote sensor 遥感器

rendezvous docking (RVD) 交会对接

repeatability 重复(合)性,重复精度

repetitive control 重复控制

resampling 重采样

rescue robot 救援机器人

residual 剩余误差,残差

resilience 恢复力,弹力,顺应力

resonance 共振

resonant frequency 共振频率

restricted Boltzmann machine (RBM) 受限玻尔兹曼机

restricted space 受限空间,限定空间

retro-reflective marker 反光标记

revolute joint 旋转关节,转动副

reward function 奖励函数,回报函数

RF (radio frequency) 射频

RFID (radio frequency identification) 射频识别
RGB (red, green, blue) 红绿蓝
RGB-D 红绿蓝-深度
RGB-D camera 红绿蓝-深度相机,RGB-D相机
right-handed coordinate frame 右手坐标系
rigid-body 刚体
rigid-body assumption 刚体假设
rigid-body displacement 刚体位移
rigid-body dynamics 刚体动力学
rigid-body model 刚体模型
rigid motion 刚性运动
rigid robot 刚性机器人
rigid transformation 刚性变换
RIGS (rate-integrating gyroscope) 速率积分陀螺仪
ring laser gyroscope (RLG) 环形激光陀螺仪
RL (reinforcement learning) 强化学习
RLG (ring laser gyroscope) 环形激光陀螺仪
RMS (root-mean-square) 均方根,均方根值
RNEA (recursive Newton-Euler algorithm) 迭代牛顿-欧拉算法
RNN (recurrent neural network) 递归神经网络
roboethics 机器人伦理学
robot 机器人
robot arm 机器人臂
robot combat 机器人格斗
robot companion 机器人伙伴
robot cooperation 机器人合作
robot gripper 机器人夹持器
robotic arm 机器人手臂
robotic exoskeleton 机器人外骨骼
robotic kitchen 机器人厨房
robotic leg 机器人腿
robotic spacecraft 机器人航天器
robotic workstation (RWS) 机器人化工作站
robotics simulator 机器人学仿真器
robot kinematics 机器人运动学
robot language 机器人语言
robot learning 机器人学习
robot navigation 机器人导航
robot operating system (ROS) 机器人操作系统
robot peripheral 机器人外围设备,机器人外设
robot programming language 机器人编程语言
robot restaurant 机器人餐厅
robot telescope 机器人望远镜
robust control 鲁棒控制
ROI (region of interest) 感兴趣

区域

rolling contact 滚动接触

rolling contact theory 滚动接触理论

root locus method 根轨迹法

root-mean-square (RMS) 均方根，均方根值

ROS (robot operating system) 机器人操作系统

ROT (remotely operated tool) 遥控操作工具

rotary joint 旋转关节

rotary motion 旋转运动

rotary motor 旋转电机

rotary optical encoder 旋转光学编码器

rotate vector (RV) reducer 旋转矢量减速器，RV 减速器

rotation 旋转

rotational inertia 转动惯量，惯性矩

rotational joint 转动关节

rotational operator 旋转算子

rotational speed 旋转速度，转速

rotational stiffness 转动刚度

rotational symmetry 旋转对称性

rotation axis 旋转轴

rotation matrix 旋转矩阵

roughing 粗加工，粗选

roughness 粗糙度

ROV (remotely operated vehicle) 遥控无人潜水器，水下机器人

RPV (remotely piloted vehicle) 遥控飞行器

RRT (rapidly-exploring random tree) 快速搜索随机树

RV (rotate vector) 旋转矢量

RVD (rendezvous docking) 交会对接

RWS (robotic workstation) 机器人化工作站

S

safeguarded space 安全防护空间
safe state 安全状态
safety reliability 安全可靠性
safety switch 安全开关
safety technology 安全技术
sample-and-hold device 采样保持器
sampled-data control system 采样数据控制系统
sampling 采样,样本抽取,抽样,取样
sampling frequency 采样频率
sampling theorem 采样定理
SAN (semi-autonomous navigation) 半自主导航
SAN (simulated annealing) 模拟退火算法
SAR (socially assistive robotics) 社会辅助机器人学[技术]
SAR (synthetic aperture radar) 合成孔径雷达
satellite navigation 卫星导航
SBL (short-baseline) 短基线
SBO (simulation-based optimization) 基于仿真的优化
scale factor 尺度因子
scan matching 扫描匹配
SCARA (selective compliance assembly robot arm) 选择性柔顺装配机器人臂,SCARA 机械臂
screw joint 螺旋关节,螺旋副
screw motion 螺旋运动
scroll chuck 三爪自动定心卡盘
SEA (serial elastic actuation) 串联弹性驱动
search tree 搜索树
second-order dynamic system 二阶动态系统
second-order linear system 二阶线性系统
second-order nonlinear system 二阶非线性系统
sectional architecture 拼合架构
security 安全(性)
security assessment 安全评估
security detector 安全检测器

security robot 安防机器人
segmentation of human motion 人体运动分割
SEIF (sparse extended information filter) 稀疏扩展信息滤波
selection matrix 选择矩阵
selective compliance assembly robot arm (SCARA) 选择性柔顺装配机器人臂,SCARA 机械臂
selective filter 选频滤波器
self-assembling system 自组装系统
self-balancing system 自平衡系统
self-calibration 自标定[校准]
self-configurable system 自配置系统
self-correcting 自校正(的)
self-coupling 自耦合(的)
self-damping 自阻尼(的)
self-evolution 自演化
self-feedback 自反馈
self-feeder 自馈器
self-healing 自修复
self-holding 自保持(的)
self-hunting 自动寻线(的)
self-learning 自学习
self-localization 自定位
self-lock 自锁
self-optimizing control 自优化控制
self-organization 自组织
self-organizing feature map 自组织特征映射
self-organizing network 自组织网络
self-organizing system 自组织系统
self-reconfigurable 可自重构的
self-reconfiguring modular robot 自重构模块化机器人
self-recovery 自修复(的)
self-relocatable 可自重定位的
self-repairing 自修复
self-replication 自我复制
self-tuning 自调节(的),自整定(的)
self-tuning regulator (STR) 自校正调节器
semantic 语义的
semantic data model 语义数据模型
semantic interpretation 语义解释
semantic map 语义地图
semantic memory 语义记忆
semantic network 语义网络
semantic recognition 语义识别
semantics 语义,语义学
semi-automatic control 半自动控制
semi-autonomous navigation (SAN) 半自主导航
semiconductor strain gauge 半导体应变计
semiglobal matching 半全局匹配
semi-gripped 半夹持

semi-Markov decision process (SMDP) 半马尔可夫决策过程
semisupervised learning 半监督学习
sense-plan-act (SPA) paradigm 传感-规划-执行范式
sense unit 传感元件
sensing coverage 传感覆盖
sensing device 传感装置
sensing element 传感元件
sensing function 感知功能
sensing uncertainty 感知不确定性
sensing-via-manipulation 通过操作的感知
sensitivity 灵敏度
sensor 传感器
sensor aliasing 传感器混叠
sensor-based control 基于传感器的控制
sensor-based controller 基于传感器的控制器
sensor calibration 传感器标定[校准]
sensor coordinate system 传感器坐标系
sensor fusion 传感器融合
sensorimotor control 感觉运动控制
sensorimotor learning 感觉运动学习
sensor integration 传感器整合,传感器集成
sensor model 传感器模型
sensor network 传感器网络
sensor noise 传感器噪声
sensor protection 传感器保护装置
sensory control 传感控制
sensory information 传感信息
sensory-motor coordination 感知-运动协调
sensory system 传感系统
sentient computing 感知计算
sequential control 顺序控制,序贯控制
serial chain 串联链
serial elastic actuation (SEA) 串联弹性驱动
serial kinematic chain 串联式运动链
serial-link robot 串联连杆机器人
serial robot 串联机器人
series elastic actuator (SEA) 串联弹性驱动器
serpentine robot 蛇形机器人
service robot 服务机器人
servo actuator 伺服驱动器
servo component 伺服元件
servo control 伺服控制
servo driver 伺服驱动器
servo error 伺服误差
servoing 伺服
servo manipulator 伺服机械臂

servo mechanism 伺服机构
servo motion system 伺服运动系统
servo motor 伺服电机
servo unit 伺服单元
servo valve 伺服阀
set-point 设定点，设定值
set-point regulation 设定点调节
sex robot 性机器人
SfM（structure from motion） 运动恢复结构
SGD（stochastic gradient descent） 随机梯度下降
shaft drive 轴传动
shallow learning 浅层学习
Shannon entropy 香农熵
Shannon's information theory 香农信息论
shape deposition manufacturing 形状沉积制造
shape matching 形状匹配
shape-memory alloy（SMA） 形状记忆合金
shape-memory alloy（SMA）actuator 形状记忆合金驱动器
shape reconstruction 形状重建
shared control 共享控制
shared manipulation control 共享操作控制
shear stress 剪应力
Shepard's interpolation 谢帕德插值
short-baseline（SBL）system 短基线系统
shortest path 最短路径
short-range navigation 近程导航
short-time rating 短时额定
shotcrete manipulator 混凝土喷射机械臂
shoulder joint 肩关节
shuttle remote manipulator system（SRMS） 航天飞机遥控机械臂系统
side slip 侧滑
Sigmoid function S型函数
signal-to-noise ratio 信噪比
simple harmonic vibration 简谐振动
simplex method 单纯形法
simplex transmission 单工传输
simulated annealing（SA） 模拟退火算法
simulated environment 仿真环境
simulated program 仿真程序
simulation 仿真
simulation-based control 基于仿真的控制
simulation-based optimization（SBO） 基于仿真的优化
simulation test 仿真试验

S

simulation theory 仿真理论
simulation to reality transfer problem 仿真到现实转移问题
simulator 仿真器
simultaneous localization and mapping (SLAM) 同时定位与建图,即时定位与地图构建,同步定位与地图构建
simultaneous motion 联动,同时运动
single-degree-of-freedom system 单自由度系统
single-input single-output (SISO) system 单输入单输出系统
single-leg robot 单腿机器人
single-master multiple-slave (SMMS) 单主机多从机,单主多从
single-purpose robot 专用机器人
single-query planner 单询规划器
single-robot SLAM 单机器人同时定位与建图,单机器人 SLAM
single-robot task 单机器人任务
single-task robot 单任务机器人
singular configuration 奇异构型
singular dot 奇异点
singular perturbation approach 奇异摄动法
singular perturbation control 奇异摄动控制
singular perturbation model 奇异摄动模型
singular value decomposition (SVD) 奇异值分解
singularity 奇异,奇异性
singularity analysis 奇异分析
singularity avoidance 奇异规避
sinusoidal locomotion 正弦移动
SISO (single-input single-output) 单输入单输出
situated learning 情景学习
situatedness 情景性
situated robotics 情境机器人学[技术]
size tolerance 尺寸公差
skating robot 滑冰机器人
skeleton computation 骨架计算
skeleton model 骨架模型
skiing robot 滑雪机器人
skill acquisition 技能获取
skill learning 技能学习
skill modelling 技能建模
skill primitive 技能基元
SLAM (simultaneous localization and mapping) 同时定位与建图,即时定位与地图构建,同步定位与地图构建
slave side 从动端
slider-crank mechanism 曲柄滑块机构
sliding constraint 滑动约束

sliding joint 滑动关节
SLIP (spring-loaded inverted pendulum) 弹簧倒立摆模型
slither 滑行
SMA (shape-memory alloy) 形状记忆合金
small gain theorem 小增益定理
smart home 智能家居
smart prosthesis 智能假肢，智能义肢
smart suitcase 智能行李箱
smart wheelchair 智能轮椅
SMDP (semi-Markov decision process) 半马尔可夫决策过程
SMMS (single-master multiple-slave) 单主机多从机，单主多从
smoothed image 平滑图像
snake-arm robot 蛇臂机器人
snake-like robot 类蛇机器人
snake robot (snakebot) 蛇形机器人
Sobel edge 索贝尔边缘
Sobel operator 索贝尔算子
soccer robot 足球机器人
sociable robot (可)社交机器人
social acceptability 社会接受度
social cognition 社会认知
social-emotional intelligence 社会-情感智能
social interaction 社会交互
socially assistive robot 社会辅助机器人
socially assistive robotics (SAR) 社会辅助机器人学[技术]
social robot 社会机器人
social robotics 社会机器人学[技术]
soft actuator 软执行器，软驱动器
soft computing 软计算
soft exoskeleton 软体外骨骼
soft finger contact 软指接触
soft gripper 软夹持器
soft material robotics 软材料机器人学[技术]
soft robot 软体机器人
soft robotic fish 软体机器鱼
soft robotics 软体机器人学[技术]
soft sensor 软传感器，虚拟传感器
solenoid valve 电磁阀
solid mechanics 固体力学
solid-state laser 固体激光器
sonar 声呐
sonar scanning 声呐扫描
SPA (sense-plan-act) 传感-规划-执行
spacecraft 飞行器，航天器，宇宙飞船
space complexity 空间复杂度
space environment 空间环境
space exploration 空间探索
space robot 空间机器人
space robotics 空间机器人学[技术]

S

space station remote manipulator system 空间站遥控机械臂系统

spanning tree 生成树

spanning tree algorithm 生成树算法

sparse extended information filter (SEIF) 稀疏扩展信息滤波

spatial clustering 空间聚类

spatial cognition 空间认知

spatial constraint 空间约束

spatial coordinate system 空间坐标系

spatial domain 空间域

spatial filtering image enhancing 空间滤波图像增强

spatial pyramid matching (SPM) 空间金字塔匹配

spatial resolution 空间分辨率

spatial semantic hierarchy 空间语义层次结构

spatial warping 空间扭曲

speaker identification and verification (SIV) 说话人身份识别和验证

speaker recognition 说话人识别

specialized robot 专用机器人；特种机器人

speech processing 语音处理

speech recognition 语音识别

speech verification system 语音验证系统

speed 速度

speed changer 变速器

speed controller 速度控制器

speed factor 速度系数

speed of response 响应速度

speed of rotation 转速

speed range 速度范围

speed sensor 速度传感器

speed variator 变速器

spherical coordinate frame 球面坐标系

spherical joint 球关节，球副

spherical robot 球坐标机器人

spherical wheel 球形轮

spherical wrist 球腕关节

spiking neural network 脉动神经网络

spine robot 脊柱式机器人

spline 样条

spline interpolation 样条插补

SPM (spatial pyramid matching) 空间金字塔匹配

sports robotics 体育运动机器人学[技术]

spot welding robot 点焊机器人

spray painting robot 喷涂机器人

spring-loaded inverted pendulum (SLIP) 弹簧倒立摆模型

spring-mass model 弹簧质点模型

SPS (standard position system) 标

准定位系统

SRMS (shuttle remote manipulator system) 航天飞机遥控机械臂系统

stability 稳定性

stability analysis 稳定性分析

stability augmentation system 增稳系统

stability criterion 稳定性判据

stability margin 稳定裕度

stability measure 稳定性度量

stability problem 稳定性问题

stability theory 稳定性理论

stabilization system 镇定系统

stable system 稳定系统

stable-state probability 稳态概率

stalling speed 失速速度

stance phase 支撑相

standard deviation 标准差

standard position system (SPS) 标准定位系统

state estimation 状态估计

state feedback 状态反馈

state machine 状态机

state observer 状态观测器

state space 状态空间

state space equation 状态空间方程

state transition function 状态转移函数

state variable 状态变量

state vector 状态向量

static analysis 静力学分析

static compliance 静态柔顺性

static equilibrium 静力平衡

static feedback linearization 静态反馈线性化

static force experiment 静力实验

static friction coefficient 静摩擦系数

static gait 静步态

static sensor network 静态传感器网络

static stability 静态稳定性

statics 静力学

statistical inference 统计推理,统计推断

statistical learning 统计学习

statistical pattern recognition 统计模式识别

steady-state 稳态

steady-state error 稳态误差

steady-state optimization 稳态优化

steady-state vibration 稳态振动

step motor 步进电机

stereo-correspondence 立体(视觉)匹配

stereo-photometry 立体光度测量,立体测光法

stereoplotter 立体测(绘)图仪

stereopsis 立体影像

stereoscope 立体视镜
stereoscopic photography 立体摄影
stereoscopic view 立体视角
stereo-triangulation 立体三角测量
stereovision 立体视觉
stereovision system 立体视觉系统
Stewart mechanism 斯图尔特机构
stiffness 刚度
stiffness matrix 刚度矩阵
stigmergy 媒介质，间接通信，共识主动性
stochastic control 随机控制
stochastic dominance 随机占优
stochastic estimation 随机估计
stochastic gradient descent (SGD) 随机梯度下降
stop-point 停止点
STR (self-tuning regulator) 自校正调节器
strain gage 应变计
stripe projection 条纹投影
strong artificial intelligence (AI) 强人工智能
structural resonance 结构共振
structured environment 结构化环境
structured light 结构光
structured light vision measurement 结构光视觉测量
structure from motion (SfM) 运动中恢复结构
subsumption architecture 包容结构
supervised learning 监督学习
supervisory control 监督控制
supervisory control system 监督控制系统
support vector machine (SVM) 支持向量机
surface fitting 表面拟合
surface modeling 表面造型
surface property 表面性质
surface stress 表面应力
surgical assistance system 手术辅助系统
surgical navigation 手术导航
surgical robot 手术机器人
surveillance and security robot 监控与安保机器人
suspension mechanism 悬挂机构
suspension system 悬挂系统
SVD (singular value decomposition) 奇异值分解
SVM (support vector machine) 支持向量机
swarming behavior 群集行为
swarm intelligence 群体智能
swarm robot system 群机器人系统
swarm robotics 群机器人学[技术]
Swedish wheel 瑞典轮
swing phase 摆动相
switching system 切换系统

S

symbol grounding problem 符号接地问题

symbolic reasoning 符号推理

synchro-drive 同步驱动

synchronization 同步

synchronization control 同步控制

synergy 协同,协同学

synthetic aperture radar (SAR) 合成孔径雷达

system architecture design 系统架构设计

systematic error 系统性误差

systematic fault 系统性故障

system complexity 系统复杂性,系统复杂度

system diagram 系统框图

system dynamics 系统动力学

system function 系统函数;系统功能

system identification 系统辨识

system integration 系统集成

system modeling 系统建模

system optimization 系统优化

system stability 系统稳定性

system transfer function 系统传递函数

T

TA (task allocation) 任务分配
tactile 触觉的
tactile array 触觉阵列
tactile feedback 触觉反馈
tactile information processing 触觉信息处理
tactile interaction 触觉交互
tactile localization 触觉定位
tactile perception 触觉感知
tactile sensing 触觉感知
tactile sensor 触觉传感器
tangent force 切向力
tangent plane 切面
tangent space 切空间
tangential distortion 切向畸变
tangential force 切向力
tangential velocity 切向速度
target acquisition 目标捕获，目标搜寻
target identification 目标辨识
target seeking 目标搜索
target tracking 目标跟踪
task allocation (TA) 任务分配
task consistency 任务一致性
task-level programming language 任务级编程语言
task-oriented grasping 面向任务的抓取
task planning 任务规划
task space 任务空间
task-space regulation 任务空间调节
task wrench space (TWS) 任务力旋量空间
taught point 示教点
TCP (tool center point) 工具中心点
teach 示教
teach and playback manipulator 示教再现机械臂
teach by showing method 示教方法
teacher 示教员
teach pendant 示教盒，示教器
teach programming 示教编程
team leader 团队领航者
telecare 远程监护
telecommand equipment 遥控设备
telemanipulation 遥操作

telemeter 遥测计
telemetering 遥测
telemetry 遥测技术
telemetry and remote control system 遥测遥控系统
telemetry simulator 遥测模拟器
teleoperated robotics 遥操作机器人技术
teleoperated system 遥操作系统
telephoto lens 远摄镜头
telephoto ratio 远摄比
telepresence 遥在
telepresence system 遥在系统
tele-rehabilitation service 远程康复服务
telerobot 远程机器人
telescopic manipulator 伸缩式机械臂
telestereoscope 立体望远镜
telesurgery 远程手术
temperature control 温度控制
temperature controller 温度控制器
temperature sensitivity 温度灵敏性
temperature sensor 温度传感器
temperature tolerance 温度耐受性
template matching 模板匹配
temporal analysis 时序分析
temporal difference learning 时间差分学习
tendon-driven 肌腱驱动
tendon tension sensor 肌腱张力传感器
tensile force 张力
tension sensor 拉力传感器
tensor 张量
tensoresistance 张致电阻效应
terminal 终端
terrain aided navigation 地形辅助导航
terrain mapping 地形建图
terramechanics 地面力学
terrestrial radio navigation system 地面无线电导航系统
test data 测试数据,试验数据
test model 测试模型,试验模型
test platform 测试平台,试验平台
tetrahedron 四面体
texture 纹理
texture mapping 纹理映射
themosistor 调温器
themoswitch 热敏开关
theory of large-scale system 大系统理论
theory of probability 概率论
therapy robot 治疗机器人
thermal actuator 热执行器,热驱动器
thermal conductivity 热传导性,导热系数,热传导系数
thermal error 热误差

thermal noise 热噪声
thermal sensor 热敏传感器
thermo-couple 热电偶
thermoregulator 温度调节器
thermoresisitance 热变电阻
thermosensor 热敏元件
three laws of robotics 机器人三定律
three-axial accelerometer 三轴向加速度计
three-axis control 三轴控制
three-axis rate gyroscope 三轴速度陀螺仪
three-axis vibrational control 三轴振动控制
three-dimensional curve 三维曲线
three-dimensional elasticity theory 三维弹性理论
three-dimensional Euclidean space 三维欧几里得空间
three-dimensional map 三维地图
three-dimensional reconstruction 三维重建
three-dimensional scanner 三维扫描仪
three-dimensional space 三维空间
three-dimensional stress 三向应力
three-phase alternating current (AC) motor 三相交流电机
three-phase asynchronous motor 三相异步电机
threshold 阈值
threshold operator 阈值算子
threshold signal 阈信号
thruster 推进器,推力器
tiered architecture 分层结构
tightly coupled task 紧耦合任务
time complexity 时间复杂度
time constant 时间常数
time constraint 时间约束
time-critical 时间关键的
time-delay 时延
time-dependent system 时变系统
time-division telemetry system 时分制遥测系统
time interval 时间间隔
time-invariant system 时不变系统,定常系统
time-of-flight (TOF) 飞行时间
time-optimal 时间最优
time response 时间响应
time scale 时间尺度
time scale integration 时间尺度集成(原理)
time sequence 时间序列,时序
time series analysis 时间序列分析
time-varying parameter 时变参数
time-varying system 时变系统
time window 时间窗
timing sequence 时标序列

TMM (transfer matrix method) 传递矩阵法
TOF (time-of-flight) 飞行时间
tolerance clearance 间隙公差
tolerance deviation 容许(允许)偏差
tone matching 声调匹配
tool center point (TCP) 工具中心点
tool coordinate system 工具坐标系
tool frame 工具坐标系
top-down development 自顶向下开发
topological configuration 拓扑构型
topological map 拓扑地图
topology 拓扑 学,拓扑
topology control 拓扑控制
torque 转矩,力矩
torque amplifier 转矩放大器
torque balance 转矩平衡
torque converter 转矩转换器
torque moment 转矩
torque motor 力矩电机,转矩电机
torque ripple 转矩波动
torque sensor 转矩传感器
torquemeter 转矩(测量)仪
torsion 扭转;扭力
torsional moment 扭矩
torsional stiffness 扭转刚度
torsional vibration 扭转振动
tracking rate 跟踪率
tracking system 跟踪系统
training example 训练样例
trajectory 轨迹
trajectory planning 轨迹规划
transducer 传感器,转换器,换能器
transfer function 传递函数
transfer matrix method (TMM) 传递矩阵法
transformation 变换
transient analysis 瞬态分析
transient behavior 瞬态行为
transient state 瞬态
transition matrix 转移矩阵,过渡矩阵
transitivity 传递性
translation group 平移群
translational frame 平移坐标系
translational mapping 平移映射
translational operator 平移算子
transmission 传输;传动,传送
transmission delay 传输延迟
transmission diagram 传动图
transmission rate 传送率,传输比
transmission ratio 传动比
transmission shaft 传动轴
transmitter 发射器,传送器
transmitting element 传送部件
transparency feedback 透明性反馈
transversal magnification 横向放

大率

trapezoidal trajectory 梯形轨迹

traversability cost 可通行性代价

treble gear ratio 三联齿轮速比

triangle generator 三角波发生器，三角形脉冲发生器

triangulation 三角定位，三角测量；三角剖分

trifocal tensor 三焦（点）张量

true random 真随机

truncation error 截断误差

Turing machine 图灵机

Turing testing 图灵测试

twist 运动旋量

twist space 运动旋量空间

two-step controller 两位控制器

TWS (task wrench space) 任务力旋量空间

U

UAV (unmanned aerial vehicle) 无人空中飞行器,无人机

ubiquitous computing 普适计算,泛在计算

UGV (unmanned ground vehicle) 无人地面车辆

UKF (unscented Kalman filtering) 无迹卡尔曼滤波

ULE (upper limb exoskeleton) 上肢外骨骼

ultimate strength 极限强度

ultrahigh dynamic strain indicator 超高动态应变仪

ultrasonic beacon 超声波信标

ultrasonic flaw detector 超声探伤仪

ultrasound 超声(波)

ultraviolet laser 紫外(线)激光器

ultraviolet radiation 紫外辐射

ultrawide-angle lens 超广角镜头

ultrawide band (UWB) 超宽带

unbalanced moment 不平衡力矩

uncertain decision 不确定性决策

uncertain system 不确定性系统

uncertainty 不确定性

uncertainty of measurement 测量不确定性

undamped oscillation 无阻尼振荡

underactuated mechanism 欠驱动机构

underactuated system 欠驱动系统

undersampling 欠采样

underwater robot 水下机器人

uniformly asymptotically stable 一致渐近稳定

uniformly distribution 均匀分布

uniformly exponentially stable 一致指数稳定

uniformly stable 一致稳定

uninterruptible power system (UPS) 不间断电源系统

universal serial bus (USB) 通用串行总线

unmanned aerial vehicle (UAV) 无人空中飞行器,无人机

unmanned ground vehicle (UGV)

无人地面车辆

unmanned underwater vehicle (UUV) 无人水下航行器,无人水下载具,无人潜航器

unmanned vehicle 无人机,无人载具,无人车辆

unscented Kalman filtering (UKF) 无迹卡尔曼滤波

unstructured environment 非结构化环境

unsupervised learning 无监督学习

upper limb exoskeleton (ULE) 上肢外骨骼

upper limb training robot 上肢训练机器人

UPS (uninterruptible power system) 不间断电源

USB (universal serial bus) 通用串行总线

UTC (coordinated universal time) 协调世界时,世界标准时间

UWB (ultrawide band) 超宽带

V

vacuometer 真空计

vacuum cup 真空吸盘

value function 值函数，评价函数

value iteration 值迭代

VANET（vehicular ad hoc network） 车辆自组织网络

vanishing gradient problem 梯度消失问题

variable stiffness 可变刚度

variable stiffness actuator（VSA） 可变刚度驱动器

variable structure control（VSC） 变结构控制

variational approximation 变分逼近

variation calculus 变分法

vascular intervention assisted robot 血管介入辅助机器人

vascular interventional robot 血管介入式机器人

vector field histogram（VFH） 向量场直方图

vehicle detection 车辆检测

vehicle formation control 车辆编队控制

vehicle-manipulator system 车载机械臂系统

vehicle routing problem（VRP） 车辆路由问题

vehicle routing problem with time windows（VRPTW） 带时间窗的车辆路由问题

vehicle stability 车辆稳定性，载具稳定性

vehicular ad hoc network（VANET） 车辆自组织网络

velocity control 速度控制

velocity curve 速度曲线

velocity feedback 速度反馈

velocity manipulability ellipsoid 速度可操作性椭球

vertical take-off and landing（VTOL） 垂直起飞与着陆，垂直起降

VFH（vector field histogram） 向量场直方图

via point （路径）通过点

vibration 振动

vibration simulator 振动模拟器

vibration suppression control 振动

抑制控制,防振控制

vibratory type gyroscope 振动式陀螺仪

vibrotactile 振动触觉

vibrotactile actuator 振动触觉驱动器

video retrieval 视频检索

video tracking 视频跟踪

VIO (visual-inertial odometry) 视觉惯性里程计

virtual assembly 虚拟装配

virtual constraint 虚约束

virtual displacement 虚位移

virtual fixture 虚拟固定,虚拟夹具

virtual human 虚拟人

virtual joint 虚拟关节

virtual reality (VR) 虚拟现实

virtual reality modeling language (VRML) 虚拟现实建模语言

virtual sensor 虚拟传感器

virtual visual servoing 虚拟视觉伺服

virtual work 虚功

viscoelastic contact 黏弹性接触

viscosity 黏度

visibility graph 可视图

vision 视觉

vision-force control 视觉-力控制

vision guidance 视觉引导

vision guided robotic system 视觉引导的机器人系统

vision measurement 视觉测量

vision sensor 视觉传感器

visual aberration 视觉畸变

visual-based control 基于视觉的控制

visual-based navigation 基于视觉的导航

visual navigation 视觉导航

visual odometry (VO) 视觉里程计,视觉测程法

visual perception 视觉感知

visual servoing 视觉伺服

visual SLAM 视觉同时定位与建图,视觉 SLAM

visual tracking 视觉跟踪

VO (visual odometry) 视觉里程计

voiceprint recognition 声纹识别

voice recognition 语音识别

volumetric map 立体地图

Voronoi diagram 沃罗诺伊图

Voronoi partition 沃罗诺伊划分

VR (virtual reality) 虚拟现实

VRP (vehicle routing problem) 车辆路由问题

VRPTW (vehicle routing problem with time windows) 带时间窗的车辆路由问题

VSA (variable stiffness actuator) 可变刚度驱动器

VSC (variable structure control) 变结构控制

VTOL (vertical take-off and landing) 垂直起飞与着陆

W

wall-climbing robot 爬壁机器人
warehouse robot 仓储机器人
weak AI 弱人工智能
weak perspective 弱透视
wearable assistive device 可穿戴辅助设备
wearable device 可穿戴设备
wearable exoskeleton 可穿戴外骨骼
wearable monitoring device 可穿戴式监测设备
wearable robot 可穿戴机器人
wearable robotics 可穿戴机器人学[技术]
wearable therapy robot 可穿戴治疗机器人
weather-proof 抗风化的,耐风雨的
weedy robot 锄草机器人
welding manipulator 焊接机械臂
welding robot 焊接机器人
WGS (world geodetic system) 世界测地系统;世界大地坐标系统,地心坐标系
wheel 轮子
wheelchair 轮椅
wheeled robot 轮式机器人
wheel-ground contact 轮地接触
white balance 白平衡
wide-angle distortion 广角畸变(失真)
Wiener filter 维纳滤波器
Wiener filtering 维纳滤波
wire guidance 有线制导
wired network 有线网络
wire-driven parallel robot 线驱动并联机器人
wireless communication 无线通信
wireless network 无线网络
wireless sensor network (WSN) 无线传感器网络
workspace (机器人)工作空间
world coordinate system 世界坐标系,绝对坐标系
world geodetic system (WGS) 世界测地系统;世界大地坐标系统,地心坐标系
worm-like robot 类蠕虫机器人

wrench 力旋量
wrist 手腕
WSN（wireless sensor network） 无线传感器网络

yaw angle 偏航角

yawhead 偏航传感器

Young's modulus 杨氏模量

Z

zero-drift error 零点漂移误差

zero-moment point (ZMP) 零力矩点

zero-order hold 零阶保持

zero pole 零极点

zero static response 零状态响应

ZMP (zero-moment point) 零力矩点

zoom camera 变焦距相机

zoom lens 变焦镜头

zoom ratio 变焦比

Z-transform Z 变换

以数字开头的词条

2.5 - D visual servoing 2.5 维视觉伺服

2 - D system 二维系统

3 - D feature extraction 三维特征提取

3 - D grid 三维栅格

3 - D model representation 三维模型表示

3 - D point cloud data 三维点云数据

3 - D printing 三维打印,3D 打印

3 - D reconstruction 三维重构,三维重建

3 - D registration 三维配准,三维拼接